BOAT ELECTRICS

James Yates

Helmsman Books

First published in 1992 by
Helmsman Books, an imprint of
The Crowood Press Ltd
Ramsbury, Marlborough
Wiltshire SN8 2HR

This impression 1996

British Library Cataloguing in Publication Data

A catalogue record for this book is available from the British Library.

ISBN 1 85223 698 1

Picture Credits

Line-drawings by Claire Upsdale-Jones
Photographs by James Yates

Acknowledgements

The author would like to thank the following for their help and
assistance in providing information and photographs for this book:
John Bennet of Lucas Marine; Peter F. Caplen; Bernard Martin of
Lucas Rists Wiring Systems; Chris Cattrall; Andy Lee of W4 Leisure
Limited; Colin Jones; Steve Hunt of Greenacre Photographic and Bob
Archer.

Typeset by Avonset, Midsomer Norton, Bath.
Printed and bound in the UK by Redwood Books, Trowbridge, Wiltshire.

CONTENTS

PREFACE

The aim of this book is to give the reader a basic understanding of electricity and its application aboard the pleasure boat. The electrical system of a boat is one of its most vital parts; poor installation techniques and shoddy materials and fittings can cause no end of problems ranging from a lack of power and the failure of navigation lights at sea, to poor navigation instrument performance and even fires breaking out because of faulty wiring and low-grade fittings. A poorly maintained electrical system will constantly inconvenience the boat owner and may very well injure or kill him. However, it is possible for the keen sailor and DIY enthusiast to learn the skills necessary in order for him to keep his electrics in tiptop condition, and many techniques are covered in this book.

Topics dealt with include how to calculate the battery capacity and power requirements for a boat, and how to choose the correct grade of fittings from lights and switches to echo sounders and bilge pumps. Installations are covered in full from planning a new system in a new boat to replacing one in a second-hand model. Cable sizes, junction boxes, connectors and wiring procedures are all shown and copiously illustrated with many black and white photographs and easy-to-understand line drawings. Highly technical explanations have been abandoned in favour of a simple approach enabling the first time boat owner to grasp the subject matter fully. All types of pleasure boat are included from small, open sports boats to big motor cruisers and ocean-going yachts.

The book also includes chapters on choosing and fitting some of the more popular items of equipment such as VHF radios, echo sounders, logs, weather information gatherers and even in-boat entertainment; TVs, radio/cassettes, etc. A special section looks at providing mains power for running power tools and other household items normally considered luxuries aboard pleasure craft. Other chapters look at basic faultfinding and how to track down and cure radio interference and electrolytic corrosion.

1
BACK TO BASICS

Mention electricity, electrics or electronics and the words will bring a *frisson* of fear into the minds of all those unlucky enough to have grabbed the wrong end of a live cable. Mix electricity with boats which brings water into the equation – especially sea water – and you have the recipe for a dangerous cocktail – lethal in the hands of the unwary or unskilled. However, electricity and a sound knowledge of its operation and installation should be part and parcel of the enjoyment of owning a boat. If you know your boat's electrical system; the batteries, distribution, fuses, cables, fixtures and fittings, you will be in a better position to be able to effect repairs when something goes wrong, will have a knowledge of basic faultfinding and will hopefully be competent enough to fit such items as a VHF radio-telephone or radar and perform simple tasks like installing extra lights for the cabin. After reading this book you should be in a position where you will be able to rewire an old boat or install a wiring system in a new one, if you are building from scratch.

The requirements of a marine electrical system are very similar to those of the average family car; both operate from a 12 volt direct current (DC) supply and some craft even have a second system giving a 240 volt capability supplied from the shore through a transformer or by portable generator. As with a car, you also need to

The electrical system is the heart of the boat. A sound knowledge of its workings will put you in a better position to effect repairs if something goes wrong.

be able to start the engine and supply the lights, the instrumentation and the radio. The difference with marine electrics is that the environment in which the system is required to work can be very hostile to electrical equipment, so wiring and fittings need different protective criteria and require a greater degree of care and maintenance in order to protect them fully from the rigors of water, salt and damp.

What is Electricity?

So, what is electricity in its most basic form? All matter is made up of millions of tiny particles called molecules which themselves consist of two or more atoms. An atom cannot possess any surplus of either negative or positive electricity otherwise we would be able to see this as we handled the particular element made up of these atoms. If we were to look at the structure of the very simplest of atoms – the hydrogen atom – we would see that it is made up of a central core called the nucleus, consisting of a single particle called a proton. One electron revolves around this proton. The proton has a positive charge (+ ve) while the electron maintains an equal negative charge (– ve). The movement of electrons produces the electric current and the materials in which this movement occurs are called conductors.

Examples of conductors are copper (which is extensively used in electrical cables as it features 29 mobile electrons), silver, water, carbon and brass. Other non-

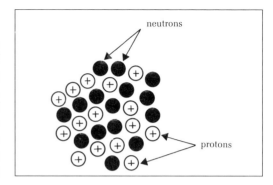

The nucleus is made up of a cluster of protons and neutrons.

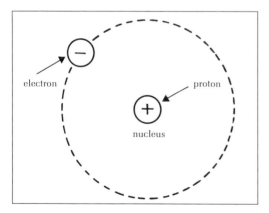

The hydrogen atom. The nucleus incorporates the proton element around which electrons orbit.

conductive materials have their electrons more closely bonded to the central atom making the passage of electrical current more difficult. We call these materials insulators and some common examples are glass, bakelite, rubber and of course, plastic which makes up the sheathing for most types of modern electrical wiring.

Electrons can be made to move from atom to atom in a controlled direction and manner along the length of a conductor. This 'flow' is caused if an action or 'force' is applied to drive it in at one end and out at the other in a circle or 'circuit'. The force is called electromotive force (EMF) and in a marine electrical system is usually supplied by a battery. If a conductor is connected across the positive and negative terminals of a source of electromotive force – such as a battery – a potential difference is created between the two ends of the conductor.

Atomic Structure

Inside the atomic structure of the conductor, the many millions of free moving electrons that have been circling their respective protons are now directly influ-

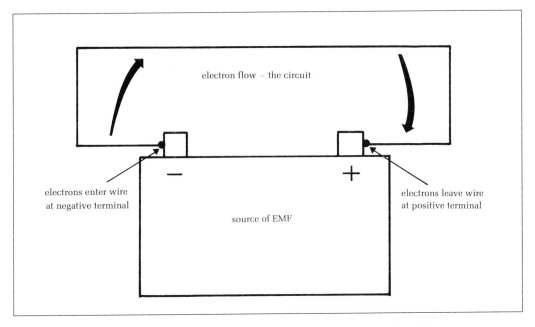

A simple circuit. The source of electromotive force – in this case a battery – pushes electrons out from the negative terminal into the conductor where they flow back to the positive terminal.

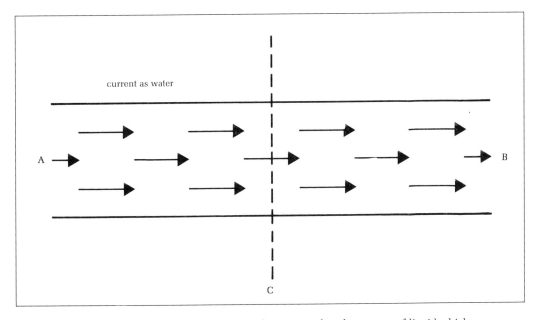

Current seen as water. The value of the current can be expressed as the amount of liquid which passes a point 'C' on its way from point 'A' to point 'B' in one second.

enced by the one controlling EMF. They are repelled by the more negative or less positive charge which has been set up at the one terminal and are attracted to the less negative or more positive charge at the other end. The abstract movement is at once stopped and an orderly flow from negative to positive occurs. This flow is called the electric current and it is measured in amperes (Amps).

The force, or pressure which is placed on the electrons from the battery is measured in Volts and the power or work that the force will eventually do is measured in Watts. The current available in this configuration depends upon the flow and the pressure behind it and is measured in Amps. The relationship between these three can be expressed as Watts = Volts × Amps (W = V × A).

Resistance

If a resistor of any sort is added into a circuit it can be likened to a brake, because it slows down the stream of electrons passing through the circuit. Resistance (R) is measured in Ohms (Ω). Looking at a wire, it is clear that the smaller the cross-section and the greater the length, the more resistance will be set up to oppose the electron flow. When this happens some of the power is lost which in turn reduces the voltage produced. This voltage drop is expressed by one of the most fundamental of all electrical formulae – Ohm's Law which states that voltage = resistance × current (V = R × A). A good maxim to remember when calculating current flow in a circuit is that current flow is both directly proportional to the applied electromotive force and inversely proportional to the resistance.

In the UK, current is always represented by the letter I, resistance by the letter R and voltage by the letter V. Of the three ways in which Ohm's Law can be expressed, the version

$$I = \frac{V}{R}$$

is the most familiar and is the one to use. The other two expressions are

$$V = I \times R \text{ and } R = \frac{V}{I}$$

As is the general rule with such formulae, if any two of the quantities are known, the other one can be found.

When an electric current flows in a conductor some heat is generated, even if it is a minute amount. A common example is the electric fire in which the element glows red as current is passed through it. The electric lamp also works on the same principle with a fine tungsten filament glowing white hot, becoming incandescent and emitting light. This heat can be put to good use as a form of safety device. For example, a wire carrying too much current for its size is a serious fire risk. But if a protective fuse is placed in line, the heat produced will cause it to melt when the current overloads the system, cutting off the power before the wiring or fitting can start to ignite. If the resistance in a circuit is too great however, the heat generated will increase to such an extent as to cause a voltage drop which in turn will effect the performance of the equipment or lighting being supplied.

Magnetic Fields

As we have already seen, when a current flows heat is generated. Magnetism is also produced by current flow, creating a magnetic field. This action is called electromagnetism and is a very important

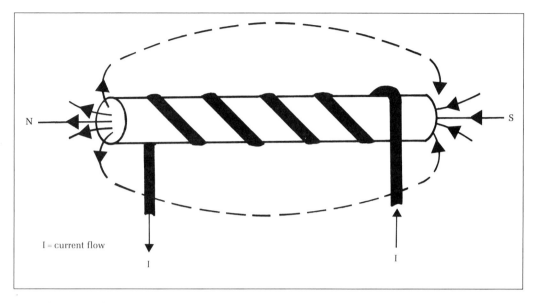

Lines of magnetic force caused by a current-carrying coil.

part of some of the electrical aspects of the boat. Motors and generators, solenoids and relays all depend on electromagnetism to work them. The magnetic force is only produced when the current is flowing but certain materials have a magnetic attraction without this current. This is called permanent magnetism and should not be confused with electromagnetism which is of a temporary nature.

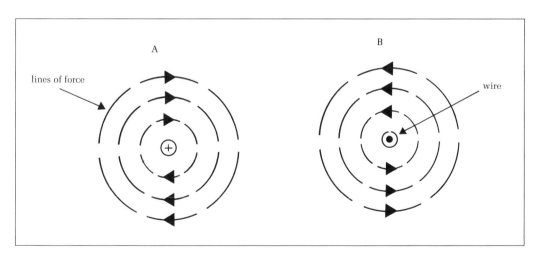

Magnetic lines of force around current-carrying conductors. At 'A' the lines go in a clockwise direction away from the viewer and anticlockwise at 'B', towards the viewer.

The effect of the magnetism is greatly increased if the wire through which the current is flowing is looped or coiled up. If a piece of iron is placed inside the coil when the current is switched on, it becomes a magnet with north and south poles. It is very important to be aware of these magnetic fields aboard the boat and to position the motors accordingly. For example a windscreen wiper motor placed near the ship's compass will influence its accuracy and could badly affect your navigational calculations. Sensitive satellite navigation or VHF radio equipment could also be affected causing crackles and interference. Electrical interference aboard a boat is a major subject which we cover thoroughly in Chapter 8. It can be traced and in most cases, suppressed.

Chemical Corrosion

Another way in which electrical current can affect parts of the boat is through a chemical form of corrosion. This is caused by minute electrical currents flowing between two dissimilar metals through the medium of the sea or in some cases, badly polluted water. All impure metals such as copper or brass contain particle pockets called anodes or cathodes (positive or negative). When immersed in a conductive medium or 'electrolyte' such as the

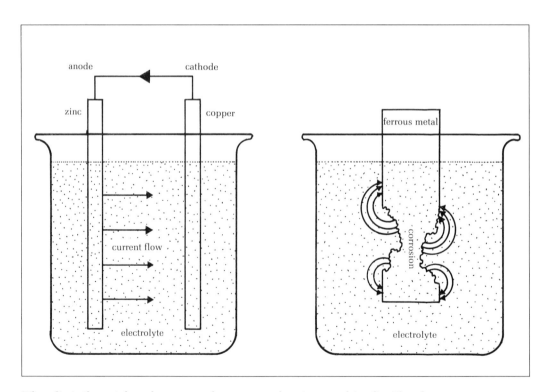

When dissimilar metals such as copper, bronze or steel are immersed in a liquid such as sea water, a small electrical battery or cell is formed. Small currents will flow from the anode to the cathode causing the anode to corrode away.

sea, a primary cell – a battery in effect – is set up causing a small current to flow from the anode or base metal to the cathode or more noble metal.

This is the same process (although much more refined) that is used during the electroplating of items of cutlery, car body shells and galvanized sheets, etc., with the result that the current takes with it minute particles of metal from the anode leaving a pitted contusion – or the corrosion that we can see. Galvanic corrosion can, like electrical interference, be traced and countered by the fitting of sacrificial anodes which will corrode away instead of the underwater propeller, stern shaft, brackets and metal fittings on the rudder. The anodes come in various shapes and sizes which enable them to be installed on the items requiring protection. For instance, small stud-shaped anodes for the rudder blade, two-piece ones for the propeller shaft and larger blocks which are fitted to the hull itself and which provide a good slab of zinc for any corrosion to eat away. Once again it is an expansive subject that is dealt with in full in Chapter 7.

Electric Shock

One final point about current; the important subject of electric shock. Electric shock to various parts of the body can be extremely dangerous, sometimes fatal, and for this reason safety should be the principal word when working on or dealing with electrical circuits or electricity in any shape or form, on or off the boat. Care should always be exercised when troubleshooting or fitting electrical wiring or equipment. Isolating switches should be in the OFF position, preferably at the battery, and fuses withdrawn when any alterations or repairs are taking place.

These then are the basics of electricity. What it is, what it does and how its various component parts – watts, volts, amps and resistance – can be calculated. We must now look at how it is actually generated aboard the boat and how to work out the power requirements for a safe and adequate supply.

SUMMARY

- In the marine electrical system the 'force' that pushes the current around the circuit is called the EMF and is supplied by the battery.

- One of the most fundamental of all electrical formulae is Ohms Law. It is usually expressed as current is equal to the voltage divided by the resistance or:

$$I = \frac{V}{R}$$

- Take great care when faultfinding or fitting new equipment into the boat's electrical system. The cocktail of water and electricity is a dangerous one – SAFETY should be the watchword.

2
THE BATTERY

There are several ways in which electricity can be supplied. The most common is your friendly local power station, another is through mechanical generators which may be either portable or fixed installation, but by far the most popular and widely used method, especially in situations such as those aboard boats, is the lead-acid battery. It could be said that it is the singular most important item aboard the boat. It is the very heart of the electrical system, and sadly, it is also one of the most neglected items of a boat's electrical fabric.

Usually tucked away in a bilge locker or deep in the engine compartment, the batteries rarely receive the attention and periodic care they so justly deserve. Not until the engine won't start one cold morning or, arriving late at night you find the lights are not working, is the poor old battery given a second thought.

This is a sad state of affairs for, as anyone who has ever had to buy one knows, a modern storage battery is an expensive item. You also have to remember that it is almost a living entity; the chemical action and particle movement inside the battery produce the electrical energy for us to tap.

The uses of batteries aboard a boat are becoming more widespread, with dual charging systems – one operating the starter motor on the engine and the other working the domestics – pumps, lights, TV and radio equipment. What is more, in these times of fuel economy and pollution control, the battery is coming into its own with a whole new generation of powered craft deriving their source of propulsion from a bank of high quality, powerful batteries.

Battery Construction

To understand our power source fully and to appreciate how important it is and how it should be cared for, it is a good idea to have some fundamental understanding of the internal workings and construction of the modern storage cell.

Most people's knowledge of a battery extends to recognizing the thing, a heavy black box with a couple of wires attached to the top from which 'magical' electricity issues. In fact it is a complex cellular and chemical factory.

Imagine two conductors (wires), one of copper and the other of zinc, are dropped into a dilute solution of sulphuric acid. If a voltmeter were put across the ends of the wires the voltage could be measured. Remember what we learnt in Chapter 1 about circuits. If the two wires were joined together outside the battery formed by the electrodes and the acid solution, the electromotive force would cause a current to flow through the wires from the copper to the zinc electrode. During this action a gas (hydrogen) is produced in the acid

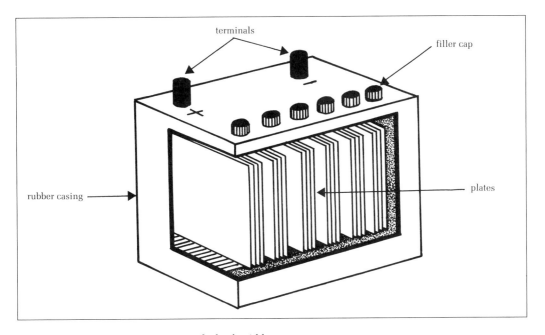

Cutaway showing basic construction of a lead-acid battery.

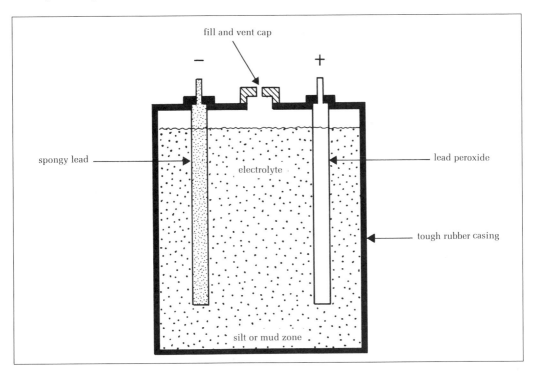

Simplified lead-acid cell.

which is attracted towards the copper electrode. This gas is replaced in the acid by zinc from the zinc electrode which will gradually become eroded. This is the simplest form of battery and is known as a cell. The copper electrode can be labelled as the positive pole and the zinc electrode, the negative pole. As we saw in Chapter 1, the current generated leaves the cell at the positive terminal and returns via the circuit to the negative one. This then is an explanation of a simple cell. Let us now put this information into the context of a commercially produced lead-acid battery.

Storage Batteries

The storage battery is an electro-mechanical device which converts chemical energy into electrical energy. The amount of power available in the battery is determined, in part, by the amount of chemical substance in the battery. Each battery is divided into a number of cells, each con-

sisting of a hard rubber compartment in which there are two kinds of lead plate – negative and positive. The plates are insulated from each other and are surrounded by a special fluid known as the electrolyte. It is the chemical action between these elements and the lead plates of the cell structure that forms and stores the electrical energy.

When the cell is fully charged, the active material on the negative plate consists of spongy lead while on the positive plate, the material is almost entirely lead peroxide.

When the cell is delivering current (discharging), the decomposition of the sulphuric acid and the formation of water submerged in a solution of sulphuric acid and water reduces the composition of both plates until eventually, a considerable amount of lead sulphate is built up. The battery should never be run down to a state where a total conversion to lead sulphate takes place or it will become unstable and will not charge.

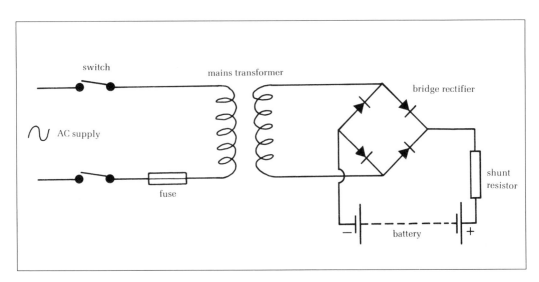

Principle of battery charging from an alternating current supply.

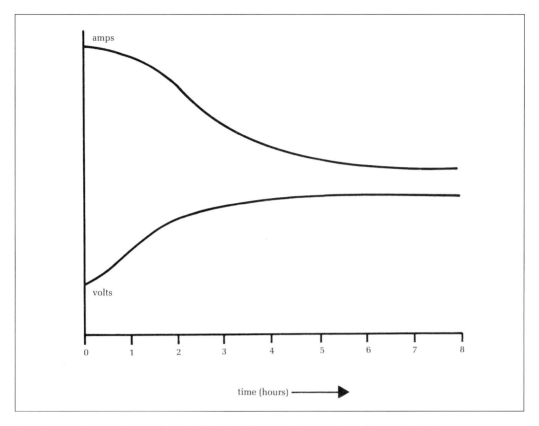

The charging characteristics of a typical lead-acid battery. The regulator allows a high charge rate at the start of charging which is gradually reduced as the battery voltage rises.

Charging

Charging takes place by connecting an external DC current across the battery terminals. This reverses the chemical action and reconverts the substances on the plates back to lead peroxide and spongy lead. If the battery is not to be used for a long period – over winter for example – it should be removed from the boat and trickle charged every 3–4 weeks with a low current input for a period of around 6 hours.

The top of each cell should also be examined and topped up as required with fresh, distilled water. Tap water is not good enough because of the chemicals it contains, nor is the so-called distilled water collected from the insides of fridges, which contains grease and other substances absorbed into it from the food. The water should be poured carefully into each cell until it covers the top of the plates to a depth of about one inch. Put in any more and you may over-dilute the electrolyte.

It is also possible to over-charge a battery. A good indication that it is reaching a fully charged condition is the appearance of bubbles in the electrolyte, or the sound of bubbling or gassing from

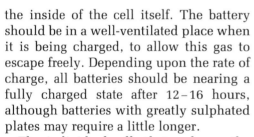

Checking the specific gravity of battery electrolyte using a hydrometer.

The level of the electrolyte should be checked and if it is found to be below the level of the plates, topped up with fresh distilled water.

the inside of the cell itself. The battery should be in a well-ventilated place when it is being charged, to allow this gas to escape freely. Depending upon the rate of charge, all batteries should be nearing a fully charged state after 12–16 hours, although batteries with greatly sulphated plates may require a little longer.

The individual cells that make up the battery can fluctuate in their ability to hold their charge. This is due to the constantly changing state of the electrolyte which is caused by the loss of acid during discharging and its replenishment during the time the battery is being re-charged. An efficient means of checking the state of

each cell is to measure its specific gravity using a hydrometer. Because acid has a greater density than water the average strength of the electrolyte can be measured by this means.

Hydrometer

A hydrometer is a simple glass tube which is weighted at one end with small lead pellets. A rubber bulb at the top end is squeezed to suck up a small amount of electrolyte after the tube has been inserted into the cell. Measurements are indicated by a small float inside the tube. The stronger the solution of acid, the higher

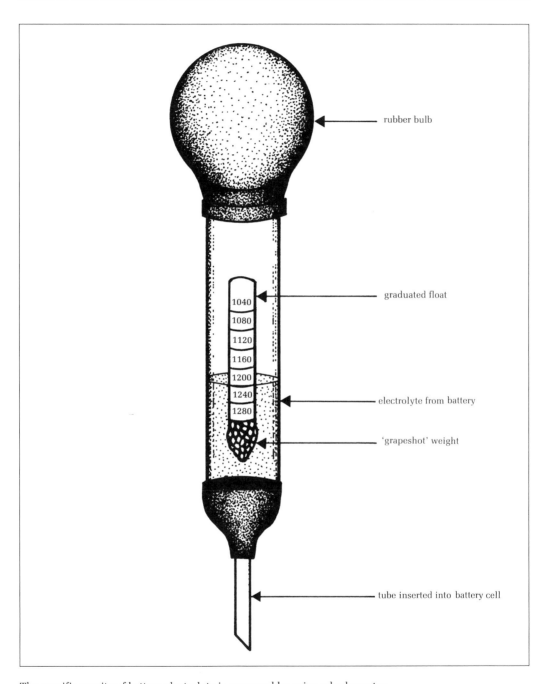

rubber bulb

graduated float

1040
1080
1120
1160
1200
1240
1280

electrolyte from battery

'grapeshot' weight

tube inserted into battery cell

The specific gravity of battery electrolyte is measured by using a hydrometer.

the float will rise; the weaker the solution, the lower the float will appear in the tube. The outside of the cylinder is marked off and the mark the float rests against will give you the specific gravity of the electrolyte in the cell. As a rough guide, a fully charged battery in good condition should have a specific gravity of about 1.25 and when fully discharged, about 1.15. Hydrometers are reasonably cheap to buy and can be obtained from your local motorists' supply shop.

On-board Charging

Obviously when the battery is in use during the boating season some form of

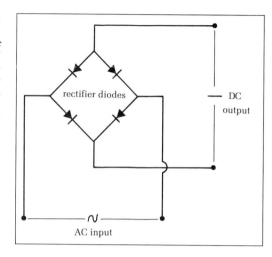

A simple bridge rectifier circuit used to convert alternating current (AC) into direct current (DC). They are used in engine alternators, battery chargers and generators.

This special rectifier can be fitted to an outboard motor. Fed with AC from a separate coil on the flywheel, it converts it to 12 volt DC with enough current to trickle charge a battery.

charging is required aboard the boat. This can be achieved in several ways. If you have an outboard motor as your propulsion source, and whether or not the battery is used to turn over the engine, many models come fitted with a charging facility. This consists of a separate coil placed in the magneto field (loop charge induction) which passes its rough AC electrical current to a rectifier. This converts the AC current to DC which is then fed, via a lead, to the battery terminals. When the engine is running, trickle charging of the battery takes place.

Alternators

With an inboard installation, the alternator or dynamo takes care of charging. The alternator produces an alternating current which is fed through a rectifier. Usually this is regulated and stabilized by the use of a 'car-type' regulator box or, more commonly now, a solid state control circuit incorporating both current and voltage control. This output is then passed to the battery in the normal way.

One final method of charging that can be used in an emergency is the 12 volt output that is now a standard feature on most portable, two-stroke petrol generators. The battery is simply connected up to the socket, and the generator run until charging has taken place.

Battery Care

The exterior of the battery should be wiped and kept clean, the securing plate (or clamp) tightened and terminals and exposed metalwork treated with water-proofing agents such as WD40 or a silicone-based grease. The same protection can be given to plug and ignition leads, distributors and coil assemblies.

One very important aspect of marine battery installation is the battery box or the area in which the battery is stowed when aboard the boat. We know that a battery is filled with acid which can be very dangerous if spilled, so it should be secured in its housing, either by fitting it into a special tray in the engine bay where it is secured with a metal clamp held down with wing nuts, or by providing a special box which encloses the entire battery.

This Hella Carri-Power battery is ideal for a pleasure boat. Its capacity is 90AH and it has been designed specifically for deep discharge and re-charge use.

These can be constructed from strong plywood or even using glass-fibre matting and resin. Remember to leave some form of ventilation in the top of the box to allow for the escape of gases which are formed during the charging process. In any case, care should be taken when working on or around the battery. Dropping a spanner across the battery terminals is a quick way of starting a fire or even blowing up the battery!

A good idea, and one that gives all-round protection and insulation of the exposed terminals, is to fit your battery into one of the proprietary plastic-cased battery boxes now on the market. These boxes can be fitted into your vented battery box on the boat and are strapped in using stout webbing and stainless steel clips. Some of these incorporate a special charging circuit to allow battery charging at home and most have large, sturdy carrying handles. Some even have the essential battery isolating switch fitted in an easy-to-reach position. As most boats are rarely within easy walking distance from the car, this helps enormously.

Maintenance-Free Batteries

Most people use a standard car or lorry battery as their power source and certain manufacturers, realizing the need for an easily portable and maintenance-free battery, have designed and built just that. Some companies supply 12 or 24 volt power sources that are re-chargeable, totally maintenance free and which incorporate useful carrying handles. Chloride Batteries for instance, have a battery ready made for leisure purposes – the Porta-power. This has a robust nylon case with a sturdy carrying handle that doubles as a terminal cover when folded down. All these, together with the new generation of solid state regulators and charging systems, now make life much easier when it comes to the important task of protecting and looking after your power plant.

Other batteries now on the market and which are eminently suitable for marine use are those which are filled with an electrolyte in a gel form. These are excellent because they are non-spill, do not produce dangerous gases when charging and should not require topping up with distilled water even after some time in use. For boats that go to sea regularly, these batteries come into their own and the convenience and safety factors involved outweigh the slightly higher cost of their purchase.

SUMMARY

- Charging of a battery takes place by connecting a DC current across the battery terminals.

- The battery should be well cared for as it is the very heart of the boat's electrical system. Without it your engine will not start, lamps will not light and equipment will be useless.

- Maintenance-free batteries are ideal for use in a marine environment. They are sealed, require no filling with distilled water and usually incorporate carrying handles.

3
How Much Power?

The Lucas manual on Marine Electrical Systems lists three basic factors which must be considered when drawing up a marine electrical system specification. They are Total Loading, Battery Capacity and Generating Equipment.

It is important to determine the capacity or size of the batteries required for your electrical installation before further planning takes place. Calculations will depend on several factors which must all be assessed before you go out and buy the batteries, switches, cable, lighting and instruments, etc.

Unlike our domestic mains supply, the electricity for a boat has to be generated on board and stored there ready to be used at any time. The mains electricity fed to our homes is AC (alternating current), but there is no way of storing this so it is relatively useless on small boats that regularly sail from port and which are away from a shore-based AC supply for any length of time. Generators can be used and are dealt with in Chapter 6, but there is really only one workable solution. A boat's supply has to be DC (direct current) which is the sort that can be stored in batteries, the operation and construction of which we discussed in the previous chapter. The electrical system depends heavily on these batteries which is why it is so important to select the correct size and type for your particular requirements.

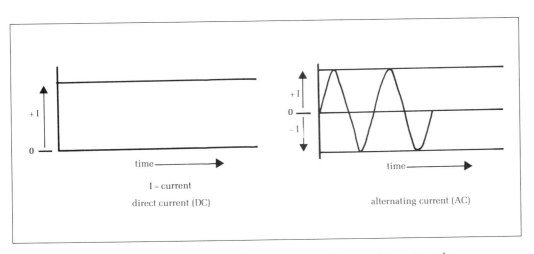

Direct and alternating currents shown in graph form against current level and time. It can be seen that AC alternates above and below zero while DC is a measured amount above zero.

Battery Capacity

The capacity of a standard lead-acid battery is directly related to the rate at which it is discharged. This figure is normally expressed as a rating of 10 hours which means that a battery having a rating of 150 amp hours will be able to supply 15 amps for 10 hours. A battery's capacity is measured by discharging it continuously at a specified rate until the voltage drops too low for any efficient use. This measurement of capacity can be used as an indication of what may be expected from any particular battery in terms of performance under load conditions.

The capacity of a battery is largely dictated by the size and thickness of its plates and how much active material is available to react with the electrolyte. We have said that the capacity of a battery is associated with the rate at which it is discharged; the lower the discharge rate, the higher the capacity and the higher the discharge rate, the lower the capacity. This is why it is important to calculate just how much is likely to be drawn from the battery in terms of current from the equipment aboard the boat.

There are two main types of battery which can be utilized aboard the boat. The first is the normal vehicle type as used in the family saloon and the second, the heavy-duty versions that are designated specially for marine use. The main differences are in the way in which the batteries are made. The car type has several very thin plates which allow it to be charged at a fast rate and then discharged to provide a heavy current sufficient to turn over the engine – its prime function. In the boat, the domestic battery, which powers all the ancillary equipment such as lights, fridge, VHF radio and radar, is required to

A small vehicle battery can be used on a day-boat for engine starting and supplying a bilge pump and perhaps a white riding light.

provide a steady current over a longer period of time between charges, something that an ordinary car battery will not do efficiently. In a car, the battery is being constantly topped up by the alternator all the time that the engine is running.

Car batteries, because of their ease of purchase, relatively low cost and light weight are often used aboard small day boats and cruisers. The strain put upon them is not too great and they can be easily taken home in the boot of the car for a quick re-charge in the garage. They will be useful for powering perhaps a small bilge pump, freshwater pump and navigation lights on a yacht or starting a small engine on a motor boat. However, for boats of any size, especially those that are regularly cruised at sea, there is no substitute for the installation of a bona fide marine battery.

If, however, a car-type battery is used it should be remembered that it will require constant topping up otherwise when used with navigation lights, radio and cabin lighting for a longish period, it may not be able to hold enough charge to get the engine started again. Slow turning over of an engine can, in certain circumstances, cause damage to the starter motor.

Twin Batteries

If several sources of power are to be supplied aboard the boat it is usual to install two batteries – one purely for starting the engine and the other for the 'domestics' – lighting, radio and navigation equipment. These are both charged from the engine alternator in the usual way, the charge current being split by an electrical circuit of blocking diodes or by a special microprocessor controlled charging monitor system which senses the state of charge in each battery and will regulate the amount provided by the alternator or other charging supply (*see* Chapter 4).

In order to determine the capacity of the batteries required for a particular boat, several factors must be taken into consideration. First, it should be decided whether a 12 or 24 volt system will be used. In the main, electrical systems in pleasure craft are usually 12 volt DC, because the initial costs of installation will be lower and the size of the batteries smaller. It is also a fact that most of the alternators fitted to marine engines are 12 volt types which tends to dictate the rest of the system. If rewiring a boat or installing a complete system into a bare hull, it might be better, perhaps, to go for a 24 volt system. One reason is that the wiring will be of a smaller gauge and switches, fuses and fittings will also be smaller in size. Most of the electronic equipment available

for marine use is available for use on either 12 or 24 volt systems so there should be no problem here. The disadvantage is that the size of the supply batteries and their weight are proportionally double what they would be for a 12 volt system.

Current Consumption

When determining battery size the likely current consumption of the boat should be assessed. If the boat is a harbour master's launch or fishing trawler that is away from harbour for long periods, the current demands on its batteries from radar, sounder and radio, etc., will be far higher than the average pleasure boat of the same size – even if fitted with similar equipment. The first thing to do is to add up the combined power requirements of all current-consuming apparatus that will be coupled to the system. This is usually expressed in watts and is done by dividing the total wattage by the system voltage. As an example, a total loading of 300 watts on a 12 volt DC system would be 300 watts divided by 12 volts = 25 amps.

A typical total loading for a medium-sized pleasure craft would be as follows:
Cabin lighting
    12 lamps at 12 watts = 144 watts
Navigation lights
    5 lights at 10 watts = 50 watts
VHF radio
    1 unit at 325 watts = 325 watts
Bilge and water pumps
    2 pumps at 50 watts = 100 watts
Other loads (navigation equipment etc)
    total 150 watts = 150 watts
The total loading would therefore be: 64.08 amps on a 12 volt system. The same loading working off a 24 volt system would require 32.04 amps.

The above is simply an example of average wattages for different items of equipment. On many boats a refrigerator will be fitted as well as an anchor winch, television and radio cassette players – some boats now fit special low-voltage microwave ovens and even such luxuries as CD, hi-fi and video, so all these requirements must be taken into consideration before arriving at your final total loading figure. It is important to calculate the average maximum total loading as this will be a deciding factor on the sizes of cables, switches and so on.

Let us look at an average small pleasure boat of 35 feet in length fitted with a 12 volt electrical system and a small 2.5 litre auxiliary diesel engine. The use of electrical equipment over an average 24-hour period might look like this:

6 × 12 watt cabin lights working for 8 hours
    576 watt/hrs
4 × 10 watt navigation lights for 6 hours
    240 watt/hrs
120 watt VHF radio used intermittently over 24 hours
    170 watt/hrs
3 watt echo sounder for 24 hours
    72 watt/hrs
Sundry small intermittent loads at 200 watts over 1 hour
    200 watt/hrs

Total power consumption = 1,258 watt hours. As the system is 12 volts the equation will be 1,258WH divided by 12V = 104AH. If the boat shares the main battery for both domestic equipment and engine starting, extra capacity must be added to include starting currents – on our 2.5 litre diesel say, 90 amp hours which will give a total battery capacity requirement of approximately 194 amp hours.

Equipment

The amount of equipment aboard will obviously vary as items are added or removed throughout the boat's useful life, and where engine starting is to be included, the starting current should be taken into consideration. As a general rule, if the engine starting load exceeds 100 amp hours, it will be best to use a separate battery just for the purpose of starting the engine. The above figures assume that the battery will be regularly charged during periods of engine running. If little charging is done during any extended period of battery use, then it will be necessary to increase the battery capacity to compensate, sometimes by as much as double the original calculated capacity.

SUMMARY

- It is important to calculate the required battery capacity for your electrical installation before any other planning takes place.

- It is usual on most boats to install a dual battery system, one battery dealing with engine starting, the other supplying all the other electrical equipment aboard.

- To calculate the average loading upon a battery, add up the wattages of lights, radio, navigation lights, pumps and other loads. Dividing the total wattage by 12 volts will give you the ampage.

4
THE CIRCUITS

The number of circuits on the boat can roughly be divided into five segments; engine and ignition, cabin lighting, navigation lights, power (fridge, TV and radio) and navigation electronics. There may also be other smaller circuits depending upon the installation. The bilge pump for example will need a circuit of its own, and will be separately switched so that it can be left 'live' to carry on working when the boat is unattended. Other circuits might be provided to supply an autopilot or electric anchor winch and will be dealt with later in this chapter.

Charging Circuit

Let us consider the charging circuit first. This circuit links the alternator on the engine to the supply batteries and is the generated source of current. The supply from this circuit feeds current to the boat's batteries which in turn distribute it to the rest of the circuits in the boat via the main fuse board. The charging circuit can be divided into two sections: one supplying current to the engine for the starter motor (if a petrol engine is used, its ignition circuit), the other supplying current directly to the batteries themselves.

The charging and distribution circuits should be isolated from each other, the only link being at the battery. The control and regulation of the charging circuit is critical for the generation of adequate supplies of current and to keep the batteries constantly topped up and ready for use. It relies on sensitive measurements of voltage and current to match the rate of charge to the present state of the battery and the loading demands put upon it. There are several devices now available that can be used to fine tune and control the current output from the alternator. They are sophisticated regulators using microchip technology which will give a visual indication if a problem with the charging circuit arises. This will tell you if more current is being drawn from the battery than is being replaced by the alternator. If this happens in practice, the solution is to either increase engine revolutions or switch something off! This

The TWC regulator senses changes in the state of charge of a battery and controls the output from the alternator accordingly.

situation can arise when using a standard voltage regulator, but the problem here is that no indication is given and even though the engine is running and the ignition warning light is out, the batteries are in fact slowly discharging. The controller should prevent this from happening by giving a warning to the owner by means of a flashing light on a panel near the helm or some other prominent place where it will be easily seen. Other problems that might cause a discharge could be a high resistance contact in the wiring harness or a badly corroded switch contact.

Regulators

The problem with standard regulators is that they normally maintain the charging voltage at a steady 14 volts regardless of the state of charge in the battery. The monitor will automatically raise this voltage to around 17 volts if necessary in extreme circumstances. This may be necessary when a battery that has been slowly discharging over a long period of time suddenly requires recharging. A standard voltage regulator would only be able to restore around 70 per cent of the batteries capacity due to plate sulphation, thus preventing the battery from receiving a full charge. In theory batteries that have a slow discharge cycle, such as those used for domestic purposes, could never be fully recharged. For instance a 100 amp hour battery will only be able to accept about 70 amp hours before a standard voltage regulator cuts off the current. The special charging monitors, although each one works in a slightly different way to its neighbour, will sense the actual state of the batteries and continue to charge until a full 100 per cent state is reached. The units

can be bought separately from marine electrical suppliers and are relatively simple to fit.

The Alternator

Most of the engines fitted to boats are marinized (that is specially prepared for use in a boat) versions of vehicle units and will be supplied with a standard alternator generating alternating current (AC). As we know from Chapter 2 on batteries and their construction, a battery can only store direct current (DC) so the AC current from the alternator has to be converted. This is done quite simply by using a rectifier comprising four silicone diodes built into the alternator casing. The alternator will supply a steady current output over the entire speed range of the engine and this is regulated from engine to engine by fitting a certain size of pulley which takes the

An alternator with the special TWC regulator brush holder fitted. The standard regulator can be seen removed alongside.

drive from the engine via a belt to the alternator. Different pulley ratios can be used depending upon the type of engine and alternator fitted to it. They should be chosen carefully to allow the alternator to supply the required amount of current for the batteries as well as preventing the engine from producing too many revs and causing damage.

Alternators are quite prone to damage and care should be taken not to disconnect the battery from the charging circuit with the alternator still working because this could cause damage to both the delicate alternator regulator and the rectifier. A switch should be fitted which disconnects the alternator's field coils before the battery can be switched off. Coupling up the battery the wrong way round is also likely to destroy the rectifier and care should be taken to check on polarity when replacing the batteries on the boat after their winter hibernation at home. A good way of doing this is to tag each cable with a positive and negative label in red and black.

Cutaway of the A127 compact marine alternator from Lucas. This high performance unit has been developed specially for marine use.

The alternator needs very little servicing apart from checking the security of its connections and making sure they are greased to keep out dirt and water. The belt tension should be checked on a regular basis as this is a prime cause of failure to charge. The best method of checking the tension is to press down on the belt at the central point between the take-off pulley on the engine and the pulley on the alternator. If the belt moves much more than an inch it should be tightened by slackening the sliding bracket attaching the alternator to the engine block and putting on more tension. Inspect the belt for signs of fraying, replacing it if worn and always carry a spare in the boat's tool-box.

Current and Voltage Monitoring

The charging circuit should also be fitted with an ammeter to monitor the flow of current for charging the batteries and a voltmeter to show the voltage being supplied. The latter is almost essential for boats with lots of electrical extras such as fridges, TVs and radios as well as the usual navigational, lighting and power requirements. The voltmeter will show the state of the batteries' voltage without the engine running and will indicate how much is being fed to them by the alternator. By judicious use of these two instruments you will soon learn to get a good idea of what is happening with the supply and demand of your electrical system.

If your boat does not already have one, it is also a good idea to fit a voltmeter. The gauge is connected across the positive and negative terminals of the main distribution board which effectively means that the meter is bridging the battery in parallel to the distribution circuits. The ammeter is connected in series between the battery and the alternator, and where heavy currents are present a shunt resistor is employed to pass the majority of the current. The ammeter is connected in parallel across the shunt and so will pass only a small amount of the total current. In this way, heavy-duty cable with its associated risk of a drop in voltage will not be required to be run up to the dashboard or control panel in order to supply the ammeter.

Another item usually incorporated into the charging circuit is a red warning lamp. This illuminates when the key is first turned to start the engine and indicates that current is being supplied from the battery to the alternator. As the engine fires up, the light goes out, indicating that the alternator has started to supply current to charge the batteries. The lamp is also a good indicator of a fault in the charging circuit. If it fails to illuminate on turning the key, it probably means that either the battery has insufficient power to turn over the engine, the alternator is faulty or there is a loose connection in the key starter circuit. If the light remains lit after the engine has started, a broken alternator drive belt should be suspected.

An excellent battery installation. Engine starter and 'domestics' batteries are enclosed in vented boxes and secured with straps. Note the position of the blocking diodes (above centre) and isolation switch (right foreground).

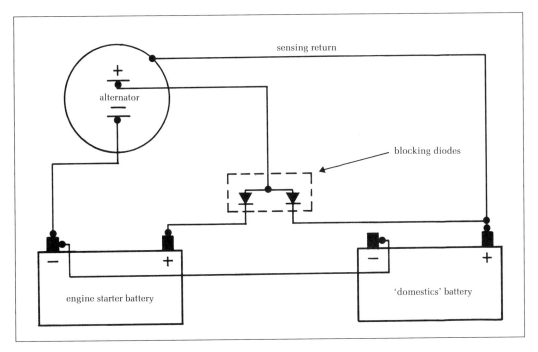

On-board battery charging using a set of blocking diodes and a single alternator.

Dual Batteries

Small boats are usually found to have only one battery supplying all the electrics as well as providing current to start the engine. This is fine if the craft is to be used on inland waters with little need for lights or other items of ancillary electrical equipment, but for boats that go to sea and regularly use many items of electrical gear aboard, there is simply no substitute for a dual battery installation – one for engine starting and the other for 'domestics'. The two batteries are linked together by a blocking diode which allows both batteries to be charged while preventing one battery from discharging into the other. This will ensure that there will always be a fully charged battery available for engine starting – a vital consideration,

especially after a heavy night aboard with extended use of lights and television!

The choice of whether to use blocking diodes or a special relay will be decided by the type of alternator fitted to your engine. If the alternator is a battery-sensed unit, it will monitor the voltage at the battery itself and the blocking diodes should be used. This can be done because the alternator output rises automatically to compensate for a voltage drop in the circuit giving a charging voltage of just over 14 volts for a 12 volt DC system. If the alternator is machine-sensed, the voltage will be controlled by the alternator's own built-in regulator which has the effect of limiting the voltage output to a level too low to compensate for the natural voltage drop in the circuit, requiring a split charge relay to be fitted.

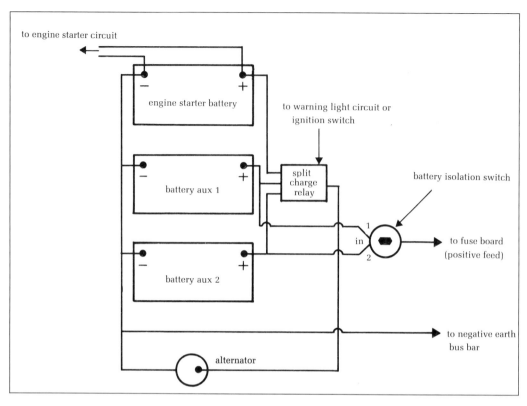

to engine starter circuit

engine starter battery

to warning light circuit or
ignition switch

split
charge
relay

battery aux 1

battery isolation switch

battery aux 2

in

to fuse board
(positive feed)

to negative earth
bus bar

alternator

A circuit for on-board battery charging employing a split charge relay and single alternator.

Engine Starting

The engine starter battery is kept solely for this purpose and should be re-charged quickly after the engine has started. The currents involved in engine starting can be as much as 500 amps and are switched via a heavy-duty solenoid. This is a quick-acting switch which is supplied with a low amperage current from the starter circuit which enables the length of the starter cables to be kept as short as possible. A good and relatively cheap way of keeping an eye on your batteries is to fit a battery state indicator. This can be either a gauge or, more likely, a row of red or green light emitting diodes (LEDs) showing a range of voltages from about 6 up to

20. The voltage zone you need to watch with a 12-volt battery is between 11 and 15 volts. If your indicator shows a voltage between 13 and 15 you can consider the battery to be fully charged and in tiptop condition. A voltage below 11 will indicate that the battery is in no fit state to do anything and must be re-charged before use.

Battery Isolating Switches

The next link in the chain of our electrical system is the distribution of the electricity from the domestic battery to the rest of the circuits in the boat. An isolation switch should be fitted in the positive line from

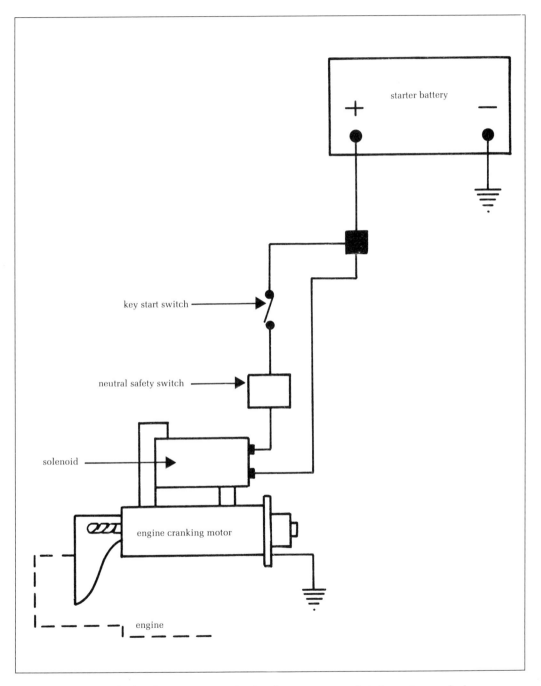

Schematic diagram of a diesel engine starter circuit using a pre-engaged starter motor, meshed with the engine flywheel.

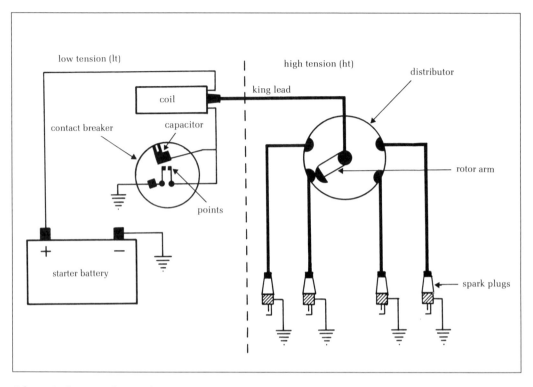

Schematic diagram of a petrol engine ignition system.

A range of Lucas battery isolating switches for 12 and 24 volt DC applications. Some versions are designed to be fitted to a battery box and some to the battery terminals themselves.

the output of the batteries. There are several types available, but the rule of thumb is to choose a good quality make with the correct current rating for your set-up. The isolator switch is an important part of the boat's electrical circuit as it is the only means of disconnecting this circuit from the power source. Both engine and domestic batteries should have separate switches and each should have a removable 'key'-type handle for added security. The switches should be mounted as close to the battery as possible – preferably to the side of the battery compartment and in a convenient position to be reached quickly in times of emergency. If a fire should break out the batteries should be among the first items to be isolated.

Fuses

The auxiliary battery supplies the distribution circuits via a distribution board and fuse panel. There are no fuses incorporated into the starter circuit because of the high current loadings, but every other circuit in the boat should be protected by individual fuses. This means that in the event of a fault occurring, the affected circuit will be instantly isolated, which should also make tracing the fault a much

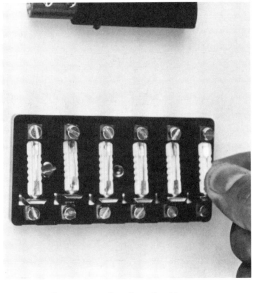

This simple, six-way fuse board with screw terminals and 1in ceramic fuses is ideal for a small craft.

Ready-made fuse and switch panels can be purchased. This six-way panel incorporates cartridge-type fuses with a separate switch for each circuit and clear labelling.

simpler task. The fuse board should be split up for each radial circuit; navigation lights, bilge pump, fridge, navigation equipment and cabin lighting. There are several ways in which the fuses can be arranged on a board. First, it would be wise to assess whether the fuses for the entire boat should be grouped together in the one place. On many boats this will be the way to go, but on some larger craft, you might decide to split the fuse boards into two; those which control lights and equipment in the aft part of the boat, and a second board situated near the helm for controlling forward lighting, navigation lights and radio, Decca navigator, etc.

Many of the craft built today have extremely sophisticated distribution boards which incorporate circuit breakers, pilot lights, switches and meters to monitor voltage and current. These are fine if you

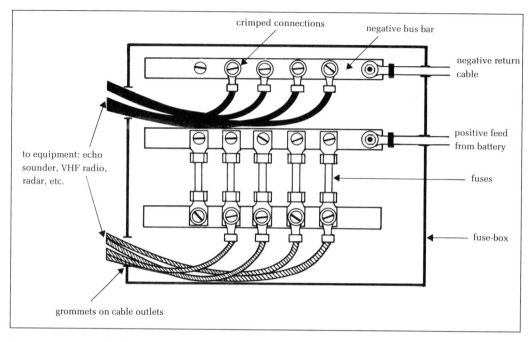

crimped connections

negative bus bar

negative return cable

positive feed from battery

to equipment: echo sounder, VHF radio, radar, etc.

fuses

fuse-box

grommets on cable outlets

Two-wire system and a five-way fuse panel. This type of panel, used on the smaller pleasure boat, uses cartridge fuses and all cables are fitted with plate connectors for good electrical continuity.

can afford them, but a simpler system can be made up by the DIY boat owner from a range of marine-grade fuse strips, lights and breakers. These are available for installation into various panels which can be labelled up neatly and are infinitely variable in their design and flexibility of use.

Mini circuit breakers are offered by several marine electrical companies. Similar to those sometimes fitted into the domestic fuse-box at home, these spring-loaded devices have a central button or switch which pops out or turns off when a fault occurs in the circuit that it is pro-tecting. They are controlled by current-sensitive coils and operate in a similar way to a normal wire fuse, cutting off the circuit when the current reaches a pre-

determined level. They are ideal for use in a boat as they remove the need for fuse changing because once a breaker is tripped it is simply a matter of pressing the button to re-set it again, after the fault has been rectified. With normal fuses, a new one will need to be inserted each time current is restored while the fault is still present. Thermal-type fuses need a few seconds to cool down before they are re-set. The circuit breaker is designed so that if a fault still exists on the line they will refuse to re-set and are impossible to jam into the ON position. You can also buy a combined switch/circuit breaker which will fit into the fuse board in place of normal switches and which performs both functions. These also make the circuitry much simpler.

A very comprehensive electrical distribution panel for the bigger boat. These are custom designed and incorporate both AC and DC circuits, volt and amp meters. They certainly look the part but are expensive.

One of the custom-made electrical panels with the door swung open. The neat installation of accessible, colour-coded wiring, circuit breakers and regulation circuitry can be clearly seen.

Fuse Ratings

When building a distribution board from scratch, make sure that the fuses or circuit breakers are rated to cover the total current load of the circuits leading to the smaller fuse panels and that the fuses at these secondary panels are rated to cover the individual loads. The panels should all be sited for ease of access and should preferably be enclosed. The usual method of mounting a fuse board is to position it in the top of a cupboard, cutting out a panel of timber, fitting the metal fuse and switch panel to it and then hinging it back into the original space. Remember to keep the input and output leads long enough to allow the panel to be swung open for service work.

When deciding upon the rating of your fuses, remember that an individual fuse or breaker should always be equal to or slightly less than, the maximum rating for any cable it is protecting. Serious fires have been caused on boats where a too high rating fuse had been fitted into a circuit having a good deal less capacity. If,

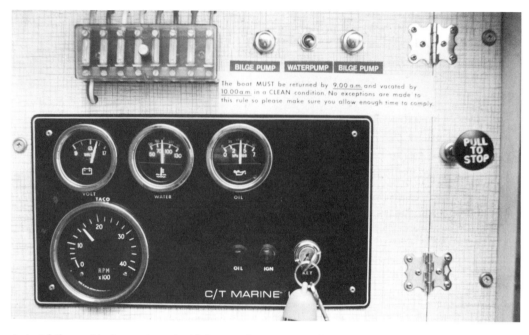

A straightforward instrument panel with battery charge state, tachometer, engine oil pressure and water temperature gauges. Oil and ignition warning lights and key starter complete the set-up with accessible fuse panel above.

for example, a lamp holder on a cabin light were to develop a short circuit, the wires feeding that circuit would rapidly start to glow red if the fuse does not blow because it had a rating that was higher than the wire itself. As many wires in boat cabins pass behind head linings and timber bulkheads, it is clear that it would not take long for a fire to start on board.

Navigation Lights

The International Regulations for Preventing Collisions at Sea are strict rules governing the positioning and type of navigation lights that must be fitted to craft. Recently these regulations were revised to include boats under 40 feet in length and to apply to all vessels cruising on the high seas and in all waters connected therewith which are navigable by sea-going vessels. The regulations do not interfere with the operation of any special rules made by the appropriate authorities for roadsteads, harbours, rivers, lakes or inland waterways connected with the sea and navigable by sea-going vessels. In addition, local inland waterway authorities may have special regulations relating to craft operating after dark, so check with any in the areas in which you intend using your boat before you start fitting the lights.

It is the responsibility of the owner, master or crew of a vessel to comply with the regulations or face the consequences of any neglect in failing to comply. Basically, the regulations cover the minimum

A set of stylish navigation lights from the Lucas range. Each lamp should have a separate circuit fed by cables of the correct current rating. Remember to keep cable runs as short as possible to prevent voltage drop.

strength of the navigation lights, measured in candelas and it is this that will determine the minimum and maximum wattages of the bulbs in each individual light. This information will be useful in assessing the rating for the cable required for wiring the lights and the fuses needed back at the panel to protect them. It is essential that before installing navigation lights approved types are purchased from a reputable supplier. The 'approved' certification will only be given after a manufacturer has had his lights tested and verified by the Department of Trade and should all be marked as being so tested. Fitting cheap and shoddy lights

could mean your insurance certificate becoming invalidated in the event of an accident at night and any claim arising therefrom.

Basic Regulations

Before going into detail about the actual fitting of the lights it is best to outline some of the basic regulations as they apply to smaller craft. 'Power Driven Vessel' means any vessel driven by machinery. 'Vessel Engaged in Fishing' means any boat with nets, lines, trawls or other gear that might restrict manoeuvrability but does not include fishing boats trolling

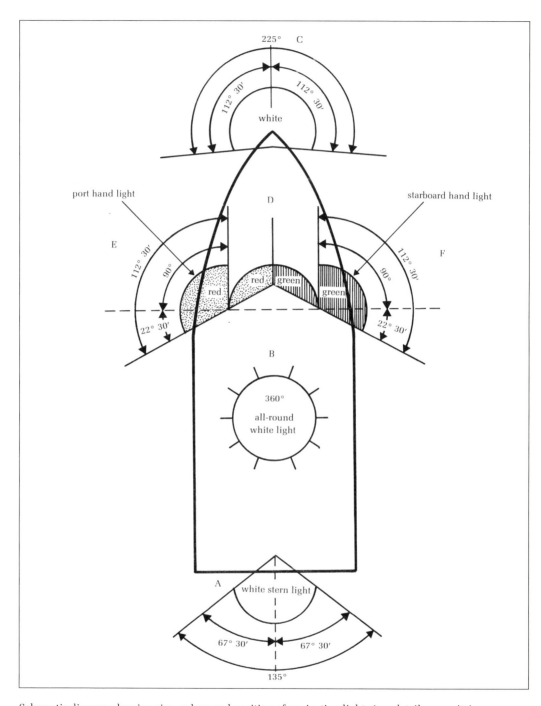

Schematic diagram showing size, colour and position of navigation lights (*see* details opposite).

NAVIGATION LIGHTS – SHAPES, COLOURS
AND POSITIONS (*see* page 38)
A: STERN LIGHT. A white light fixed at the
stern of the vessel and showing an unbroken
light through a horizontal arc of 135 degrees.
67½ degrees can be seen from the aft end of
each side of the vessel.
B: ALL-ROUND LIGHT. A white light giving an
unbroken light through a horizontal arc of 360
degrees.
C: MASTHEAD LIGHT. A white light fixed over
the fore and aft centreline of the vessel showing
an unbroken light through a horizontal arc of
225 degrees – forward from 22½ degrees abaft
the port and starboard beams.
D: BI-COLOUR LIGHT. A two-colour red and
green light fixed over the fore and aft centreline
of the vessel showing a red unbroken horizontal
arc of light from the forward centreline to 22½
degrees abaft the port beam and a green
horizontal arc of light from the forward
centreline to 22½ degrees abaft the starboard
beam.
E: PORT SIDE LIGHT. A red light showing an
unbroken arc of 112½ degrees forward from a
line parallel to the centreline of the vessel to
22½ degrees abaft the port beam.
F: STARBOARD SIDE LIGHT. A green light
showing an unbroken light through a horizontal
arc of 112½ degrees forward from a line parallel
to the centreline of the vessel to 22½ degrees
abaft the starboard beam.
TOWING LIGHT. A yellow light with the same
characteristics as the Stern Light.
HALYARD LIGHTS. White, red and green lights
that show an unbroken light through a
horizontal arc of 360 degrees and with a
minimum range of two miles.

lines. 'Vessel Not Under Command' is any
vessel which through some exceptional
circumstances is unable to manoeuvre as
required by the regulations and is therefore unable to keep out of the way of other
vessels. 'Underway' is when the vessel is
not at anchor, aground or made fast to the
shore.

The principle behind the carrying and
operating of lights is very similar to the
more familiar road vehicles – that is see
and be seen, both in darkness and in poor

visibility. Navigation lights must be
switched on from sunset to sunrise and
during those times no other light should be
shown other than those which cannot be
mistaken for the lights specified in the
regulations, or do not impair the visibility
or distinctive character of the navigation
lights. These lights can also be shown
during the hours of daylight during times of
restricted visibility or circumstances where
it is deemed necessary, such as heavy rain,
fog, mist and in rough waters where the
visibility of small boats may be difficult.

The various types of lights, their colours
and areas of illumination can be seen by
referring to the table. The horizontal angle
of arc shown by the illuminated light
should enable a boat to be able to judge the
approximate course being steered by the
vessel in sight. A boat which is coming at
you head-on will show both port (red) and
starboard (green) lights plus the masthead
or all-round white lights. A boat cruising
across your track will show either a port or
starboard light along with the masthead. If
you are coming up astern of the other boat
then you will see either one or two other
white stern or masthead lights. The only
deviation from this will be the all-round
white light shown on a boat that is under
7m (23ft) in length with a maximum speed
capability of less than 7 knots.

The second diagram shows the fitting
heights for the positioning of the lights on
the boat. These heights must be correctly
observed as to deviate from them in any
major way would confuse another boat approaching in the dark. The heights and
position of the various lights give a clear
indication as to the type of boat, its overall
length, speed and whether it is a boat
under power, sail, stopped or engaged in
fishing activities. On a vessel in the
12–20m (39–67ft) range the masthead

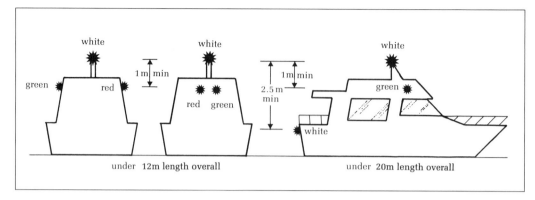

Minimum heights of navigation lights on vessels under 12m (39ft) and 20m (67ft) length overall.

light must be fitted at a height not less than 2.5m (8ft 2½in) above the gunwhale and there should also be a minimum of 1m (3ft 3in) between the height of the masthead light and the port and starboard lights. On vessels that are less than 12m (39ft) in length the masthead light can be mounted at a height less than 2.5m (8ft 2½in), but there still has to be a 1m (3ft 3in) difference between it and the port and starboard lights.

Sailing Craft

A sailing boat, even when fitted with an auxiliary engine, needs a different navigation light combination. When the boat is cruising it must show port and starboard

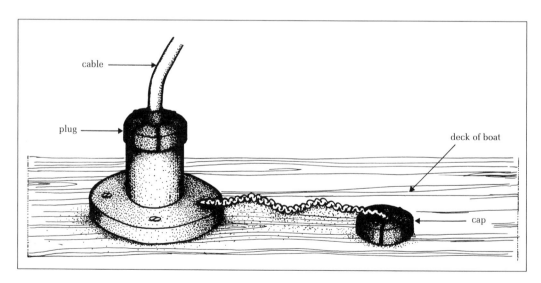

Where a cable is required to pass through the deck, a waterproof marine-grade plug should be used. The cap covers the socket connections when the plug is removed.

side lights and a white stern light. If the boat is less than 12m (39ft) in length, the two side lights can be combined together in a single unit near the top of the mast, or a three-light unit; red, green and white (tricolour) at the masthead. This single light can have just one bulb and will help to cut down on current consumption when under sail. If the yacht should start her engine, she immediately comes under the 'Vessels Under Power' regulations and should comply with the navigation light rules in that she must show a 225 degrees forward-facing white light at the masthead which should be the required 1m (3ft 3in) above the level of the port and starboard

Navigation lights can be mounted at the correct height on sea railings with their cables passing into the boat via special deck plugs, NOT through a drilled hole filled with mastic!

side lights. If the boat is less than 7m (23ft) in length, it can motor under a white all-round light plus the two side lights.

If the yacht is fitted with a three-colour masthead light this should be extinguished if the boat is under power at night. This is to avoid the confusion of another boat seeing a red and green masthead light above the red and green side lights fitted below. The white masthead riding light should also be turned off and cannot be used as a motoring light if the yacht is less than 7m (23ft) in length.

Fitting Navigation Lights

The siting of the navigation lights and the routing of cables is an important part of your electrical installation. Cable sizes should be checked with the manufacturer and only approved DoT certified lamps should be used. It is very important to use the correct size of cable for your navigation light supply. The regulations state that lights should produce a certain output depending upon the length of the boat and this output depends upon the power available at the bulb. It is no good buying the best, approved navigation lights with the correct bulbs fitted if the supply cable is of insufficient capacity to supply the lamp with enough current. Where cables come out of the deck, they should be fitted with waterproof deck glands, preferably in tough stainless steel to help them to withstand the rough and tough deckwork and rope and chain shifting that sometimes takes place. Use only marine approved glands or plugs and sockets and ensure that they are fitted using the sealing gaskets supplied. On some of the lights that are fitted to the cabin sides or on the wheel-house, cables can probably be made to come out of a simple hole, protected by

Another method of routing navigation light supply cables is to feed them into the sea rails which act as conduit. Remember to fit a rubber grommet at the point of entry to prevent chafing.

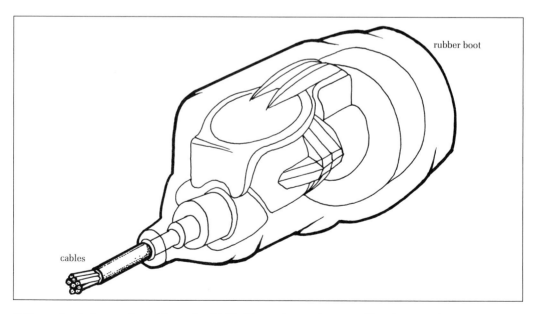

Cables and switches used outside can be shielded by enclosing them in rubber 'boots' and sealing with self-amalgamating tape.

a rubber grommet. Extra waterproofing protection should be given by smearing silicone sealant over the grommet and wire. The lamp holders and connections on the lights should be periodically checked for signs of corrosion and cleaned up as necessary. A quick spray with WD40 will keep contacts clean and help to repel water. Always carry a good stock of spare bulbs and remember to check that each light is working before leaving the marina.

Maintaining Night Vision

On some boats fitted with flybridges and on yachts where the helmsman is required to steer from outside the cabin, the position of certain navigation lights combined with their natural brightness could cause problems with night vision. Unless you have experienced a long night voyage and realize how important night vision is to keeping a keen lookout for other vessels, buoys and navigation marks, you will not believe how important it is to have shades fitted to your navigation lights. These can be fabricated from glass fibre or timber and are fitted to the portions of the lights that shine directly onto the helm position. If fitting shades, remember that no part of any light should have its designated angle of view obscured as this will be dangerous to other craft and will infringe the regulations.

Cabin Lighting

The lighting circuits on board should be doubled up to give a selection of two types of lighting: filament bulbs and fluorescent tubes. There is a much better light output from fluorescents and they do take slightly less current, depending upon the type

fitted. They do, however, cause interference with the radio and certain items of navigation equipment because of the frequencies employed in the starter circuits, which ordinary filament bulbs do not. Another sound reason for having two lighting circuits is that if one circuit should become inoperative, the other should still be available and you will not be plunged into darkness.

Lighting aboard is to some extent a matter of taste, but remember that boating can be a wet and windy activity – especially when cruising in British waters, and a good deal of time might have to be spent below decks. An adequate source of lights to brighten up cabins and worksurfaces is essential for comfortable living aboard. Most boats have some sort of lighting fitted from new but it really depends on the individual boat builder how these lights are placed and how many are fitted. If you are buying a boat from new you will probably be happy with the lighting as you will have assessed it before parting with your money, but if buying second-hand or fitting out from scratch, you will be able to plan a complete lighting arrangement custom-built to your own specification.

A good mix of both fluorescents and filament lamps is ideal. Fluorescents, because of their high light output should be used as 'blanket' lighting in the main cabin and wheel-house with smaller filament lamps being used at berths for reading, and in toilet compartments where space does not dictate the fitting of bright lights. Good lighting is required for the galley area especially as it is here that you will be preparing food, using sharp knives, handling saucepans, boiling water and hot fat. All these are potentially dangerous if you cannot see clearly what you are doing. The chart table should also be well lit, but a smaller light, perhaps one on a flexible arm, should also be installed here for night-time use when a bright light below would destroy the night vision of the navigator. It might be an idea to install a red light near the chart table which can be left on permanently during the hours of darkness to help crew members see as they come off deck whilst still retaining their night vision.

Cupboards and Steps

A good place to fit a small filament light is at the back of a large food storage cupboard or wardrobe unit which is capacious enough to need one when you darken it by standing in the doorway, blocking off the light. This light could be activated by a small door switch similar to that fitted to a domestic fridge, and would turn off the light as the door was closed. These lights need not be very big or bright, just sufficient to illuminate those dark recesses. The dining area can either be lit by one central fluorescent light, or by more subdued and discreet lights hidden behind small pelmets running around the outside area of the seating. Remember that in areas such as the toilet compartment and bathroom, steam and water splashes could affect the lights so choose units that are impervious to water and steam, especially if the boat is fitted with a shower. You will also need to fit a special shaver point in here comprising a small box with a two point plug on the front containing a transformer and circuitry that converts the boat's 12 volt DC supply into the 240 volt AC required by the average domestic shaver. These devices should always be wired in on their own separately fused circuit.

Steps, corridors and dark companionways where crew members regularly walk are difficult to distinguish in half-light and can be dangerous especially if the boat is being thrown around in a rough sea. Small bulkhead lights are available, fitted with or without individual switches, that can be mounted either just above or behind steps or low down near the cabin floor. They give out sufficient light for the steps and companionways to be seen clearly. All too often the engine room is a dark, murky hole where it is difficult to distinguish various aspects of the engine or its associated equipment. This is especially true if the engine is situated inside the cabin of the boat beneath the saloon or cockpit floor. It is not always practicable to lift up all the floorboards in order to check engine oil and water levels or to check for leaks in the stern gear, but a couple of heavy-duty bulkhead lights fitted with protective metal grills will illuminate the compartment sufficiently.

Current Consumption

Once you have planned the types and position of your lights in all cabin areas the next thing is to sit down and work out the total current consumption of each lighting circuit. Start by adding up the wattage of the lights on each circuit you intend to install, dividing this sum by the voltage (12 or 24V) and this will give you the total amps consumed each hour assuming that all lights are in use at the same time. The formula is $W = A \times V$. Each of the lighting circuits should be provided with its own fuse or circuit breaker. On larger boats more than two circuits might be required, for example in the main saloon, where one will still function if the other should 'blow'. The same idea could be applied to boats having several cabins and compartments where lights can be divided into port and starboard circuits, again leaving cabins with at least one light burning in the event of a failure on the other line.

Lights, like any other item of equipment aboard the boat should be chosen with care. Once again, you really only get what you are prepared to pay for and you should always go for the best quality fitting you can afford. There are several manufacturers producing a wide range of lights suitable for use in a marine environment and these should always be used in preference to the ones designed for domestic applications.

Bilge Pump

The bilge pump will require a separate circuit because it will need to be kept 'live' in the automatic mode whenever the boat is left afloat unattended. The choice of pumps is very wide and ranges from simple manual diaphragm types that are installed in the cockpit and worked by a hand-operated lever to powerful electric ones that can shift many gallons of water every minute. The pumps are available in either 12 or 24 volt DC versions and are normally mounted 'loose' at the lowest point of the bilge in an area where they can be accessed for servicing and cleaning. They comprise an upright pump unit where water is drawn in the base and a single hose coupled to a skin fitting on the hull above water-level ejects it over the side. Many types now have an automatic mode fitted as standard, which means that the pump can carry on working when the boat is moored up or left at anchor. This alleviates the need to remember to switch

Every boat should have at least one bilge pump fitted. It should be supplied on its own, separately fused circuit.

the pump on before leaving the boat. A float switch is incorporated which turns on the supply to the pump when the water reaches a certain level. The float switch is a mechanical tilting device that moves as the level of water increases. Sometimes these can jam up with silt and dirt, so check them when you inspect the bilges. Alternatively, you might consider fitting one of the new 'sonic' switches that work by detecting the flow of water between two sensors. These have no working parts and therefore rarely go wrong, requiring a simple clean out when necessary.

Anchor Winch

Another circuit that will need its own separately fused run will be the one that feeds the electric anchor winch, if you are going to fit one. These winches draw quite

Larger craft may be fitted with an electrically operated anchor winch. These draw a heavy current and should have supply cables of an appropriate size. This winch has a remote control facility for ease of use.

heavy currents, depending upon their type and will require special heavy-duty cables for their supply. These should be as short as possible and routed in conduit in the usual manner. A special relay can be purchased with some winches to allow the anchor to be raised and lowered automatically using a plug-in remote control on deck. This is a good idea because it means the person operating the winch can see exactly what is happening from the bow and is able to shout precise instructions to the helmsman. It is essential to have the engine running before trying to raise a deeply buried anchor. This will ensure that enough current is being pumped into the batteries in order to compensate for the heavy current draw required to break an anchor out from the mud.

SUMMARY

- The electrical system on board should be divided up into five circuits; engine, cabin lighting, navigation lighting, power and navigation electronics.

- Alternators, which supply the charge to the battery, are quite prone to damage. This can be caused by disconnecting the battery with the engine still running or by connecting it up the wrong way round.

- When building a distribution board make sure that the fuses or circuit breakers are rated to cover the total current load of all equipment likely to be used on any particular circuit.

5

INSTALLING THE SYSTEM

When it comes to the nitty gritty, getting down to the practical aspects of installing the wiring on the boat; there is really no substitute for planning your circuits and if possible drawing it all out to scale on a large piece of paper inside the boat itself. In this way you will be able to calculate the amount of cable, light fittings, fuses, panels and junction boxes required. It will also give some indication as to the final cost of your installation. The best way to start a plan is to make a scale drawing (as accurately as you can) showing a plan view of the boat. You will then be able to mark on this the relative positions of all your electrical equipment; lights, socket outlets, radio and navigation equipment positions, bilge pump and engine wiring.

You will need several separate circuits, for lighting, power and communications for example, noting which items of equipment can be allocated a shared circuit and which will require a separate one. You can decide this by calculating the power each item will require and also by deciding which items are likely to be in use at the same time. All the circuits will need to be separately fused, so you will also have to decide how many fuse boards you will require. On a large craft you may need several, depending upon the length of your cable runs.

A superb example of a neatly laid out instrument panel which combines engine functions, steering and controls, compass and switch and fuse panel. Note the echo sounder (top centre) installed so that it can be easily read by the helmsman.

Whatever layout you finally decide upon, one factor remains constant; the engine wiring and battery charging circuit must be kept separate from the domestic and equipment circuits.

Essential Links

It is the wiring aboard the boat that provides the essential links between the batteries, alternator, fuse board and the equipment to be powered, and its install-

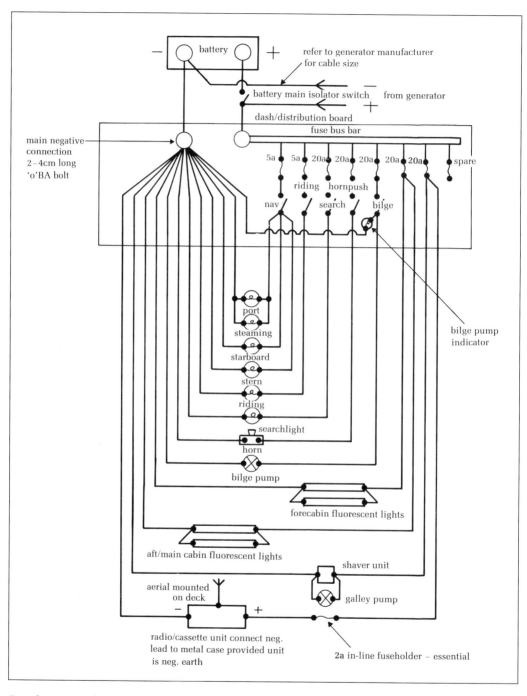

Complete wiring diagram for a small craft electrical system.

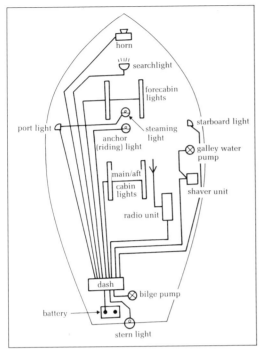

Schematic diagram for a small craft electrical system.

This electrical distribution board shows what can be accomplished when money is no object. Everything is labelled, colour coded and easy to get at making faultfinding a much simpler task.

ation is therefore vital to the safe and functional operation of all systems from engine starting to lighting and navigation. The environment afloat is extremely hostile to any form of wiring and it is only by using the correct grade of cable and a well thought out installation procedure that reliability and trouble-free service can be ensured. Most boat builders today recognize the need to provide high standards in the way their products are wired and in the pleasure boat market these have improved considerably over the last ten to fifteen years. Stringent regulations are laid down for wiring installations and it would be folly to ignore them when doing your own installation or indeed, when adding extra circuits to an established system.

Double Wiring System

All wiring circuits should be supplied with a double system comprising separate cables for both outward and return legs. This applies even to circuits involving the engine where the metal of the engine block is sometimes utilized as a return, as it often is in a vehicle. A double wire system should be used to prevent obscure current leakage faults and the corrosion that will be the result (*see* Chapter 7 covering galvanic corrosion). This is particularly important when installing an electrical system on an all-steel boat. It might be tempting to use the hull itself as the return path, but this will inevitably result in major galvanic corrosion.

You can install a 240 volt AC ring main similar to the ones in a domestic house on the boat which will be supplied via a transformer and plug/socket from a shoreside supply. It is bad practice, however, to install a DC ring main because of the magnetic field that will be set up. This could have adverse effects on the ship's compass causing quite major errors in your navigation calculations.

Cable Type

You should never use domestic twin and earth-type cable in the boat, or switches and other ancillary equipment which has

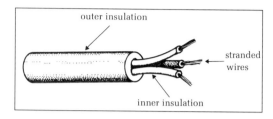

It is always better to use stranded core cable rather than single core to avoid the problems of fracture caused by movement.

been rated for AC current because current flow differs considerably between AC and DC circuits. You will remember from Chapter 1, that current is measured in amps and that this can be calculated as watts divided by volts. A 150 watt domestic bulb supplied by a 240 volt AC mains circuit will produce a loading on that circuit of 0.63 amps. Now consider an average 30 watt bulb in the 12 volt DC circuit of a boat's cabin lighting. This consumes a much larger 2.5 amps which means that the rating for the cable used has to be much greater and this is without considering any voltage drop caused by longish runs in the cable.

It is very important that your wiring should be of the correct cross-section and current rating for each circuit that you want to build into the boat, so that it has the capacity to carry the current expected of it and in order to reduce any dropping in voltage. Normal domestic wiring is made of solid conductors which might be easy to work with and are flexible and readily available in the shops, but the constant movement aboard a boat can cause these conductors to flex and fracture – sometimes in parts of the cable that can not easily be reached. This may even necessitate the replacement of an entire circuit. The wire of the cable should be stranded instead of solid which will

Tidy and ordered wiring behind the dash panel. This is the point at which many electrical cables eventually converge and should be as ordered as possible. Here extensive use has been made of plastic spiral cable wrap.

eliminate the fracturing problem as well as any possibility of a fire starting due to sparking at the break. Each of the wires must be both insulated and sheathed. Under certain regulations, for instance the very strict ship wiring ones laid down by the Lloyds A100 specification, the cables should be sheathed in two types of synthetic rubber. For normal use, though, the wires can be sheathed and insulated using PVC.

PVC sheathed cables are the most commonly used on boats that are not seeking to comply with the A100 specification, but there are different grades available, including a special heat-resistant PVC capable of handling higher current loadings and providing much better protection in areas of high heat generation such as engine compartments. It is here that heat, vibration, oil and fuel all band together to attack the wiring and installing a higher standard of cable than that used for the less critical circuits elsewhere in the boat will pay dividends in both reliability and peace of mind.

Engine Compartment Wiring

In the engine room all wiring should be clipped up off the bilge at regular intervals using proper cable clips. If possible wiring looms, or better still, conduit should be used to shield the cables from excess heat and grime, and should be considered essential for use in engine and fuel tank spaces.

On the engine itself, the big, heavy-duty battery and starter cables should receive particular care. They can carry as much as 450 amps and should therefore be kept as short as possible to prevent voltage drop. The connections to the battery terminals and alternator lugs should be of sufficient

Wiring installations in the engine compartment should be carried out to the highest standard possible. Heat, oil and fuel can all take their toll on a shoddy installation. Note how the wiring is neatly clipped together and routed well away from the engine and propeller shaft.

size to handle such large currents and should be crimped into place on the cables for maximum security. The terminals themselves should be kept as clean and bright as possible.

If, on trying to start the engine, the

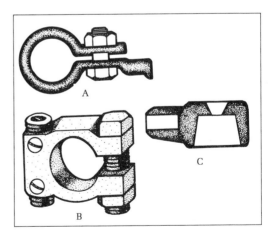

Three examples of battery terminals. A: clamp-on with soldered cable connections; B: clamp-on with screwed cable connection and C: push-fit with central securing screw.

A small electrical fan has been installed in this engine compartment to help remove fumes and petrol vapour. It is a special no-spark fan for marine use and is switched on for several minutes before starting the engine.

turnover is slow and sluggish, the fault (which might be a flat battery) could well lie in the starter cables. If cables of insufficient cross-section have been used, the drop in voltage may be enough to prevent a fast and efficient cranking of the engine. However, if terminals and cables are loose and/or dirty this can also be a prime cause of intermittent or poor starting. Check these first!

Cable Ratings

It is very important to check on the gauges of any wire which is to be used for a marine electrical circuit. Boatyards and equipment and cable manufacturers will have special booklets of tables showing cross-sections and current ratings for all the cables they stock. These should be scrutinized before buying. Cable is graded by the cross-sectional area of its conductor and is usually given in square millimetres. This cross-sectional area gives the current capacity in a similar way in which the diameter of a water pipe will determine how much water will be able to flow through it. It is particularly important to

Nominal cross-section, mm²	General purpose rubber and PVC		'High temperature' or heat resisting PVC		Butyl		Ethylene propylene rubber, cross-linked polyethylene		Res. ohms per 1,000m at 20°C
	single core amps	2 core amps	single core amps	2 core amps	single core amps	2 core amps	single core amps	2 core amps	
1	8	7	13	11	15	12	16	13	18.84
1.5	12	10	17	14	19	16	20	17	12.57
2.5	17	14	24	20	26	22	28	23	7.54
4	22	19	32	27	35	30	38	32	4.71
6	29	25	41	35	45	38	48	40	3.14
10	40	34	57	49	63	53	67	57	1.82
16	54	46	76	64	84	71	90	76	1.152
25	71	60	100	86	110	93	120	102	0.762
35	87	74	125	105	140	119	145	120	0.537
50	105	89	150	127	165	140	180	155	0.381
60	120	100	175	150	185	160	200	170	0.295
70	135	115	190	161	215	183	225	191	0.252

Table of cable ratings and sizes (Courtesy of Lucas Marine).

realize that the rating and size of a cable is based on the resistance of its conductor and the value of the temperature rise allowed in its insulation. The table gives some examples of current rating for popular cable sizes including those that should be used in a marine electrical system meeting Lloyds requirements.

Waterproofing

If a cable should pass through a watertight bulkhead a waterproof cable gland should be used and a small drip loop created in the cable to prevent the 'wicking' flow of water. Some boat circuits of 12 or 24 volt DC also have the 240 volt mains AC circuit supplied by an on-board generator or shore supply (see Chapter 6). It is essential that these cables are correctly routed and protected in order to prevent the possibi-

The wiring through this mast terminates in special deck plugs and sockets, but is fed through an unprotected hole in the aluminium mast itself. This is bound to cause chafing and eventual electrical failure.

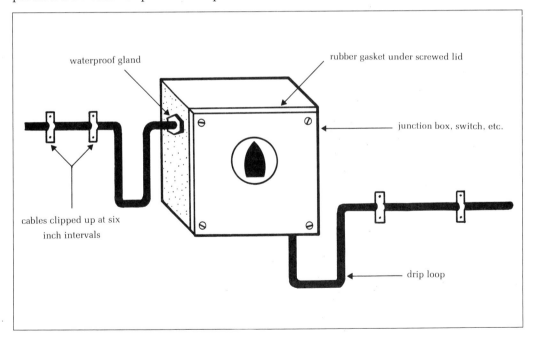

Correctly fitted exterior waterproof installation. Note the drip loops on cables and the waterproof gland where cables pass into the switch box.

The right way of doing it. These cables are fed individually through protective rubber grommets before terminating in waterproof deck sockets.

lity of electrocution or fire. All components and cables should be protected from water which means the use of sealed junction boxes and special galvanized conduit. If a cable is required to pass through an ordinary bulkhead a grommet should be inserted before the cable is fed through. Special deck glands are available in straight, or plug and socket versions for cables that have to pass through the deck – for example navigation lights.

Junctions and Connections

At some time you will need to make connections when wiring the boat or fitting extra circuits into the existing main wiring loom. It is vital that these connections are made correctly. They are just as important as the size of cable used and could develop into a potential trouble spot if not properly made. It is a great temptation to strip back the insulation from two cables to be joined and twist the ends together, 'insulating' the joint with PVC tape. This may be fine for a temporary repair to a cable carrying a light loading, but absolute disaster if left in the marine environment long term.

Waterproof deck plugs and sockets should be used whenever a cable is required to pass into or out of the boat. This one from Dri-Plug will accept up to 19 separate wires.

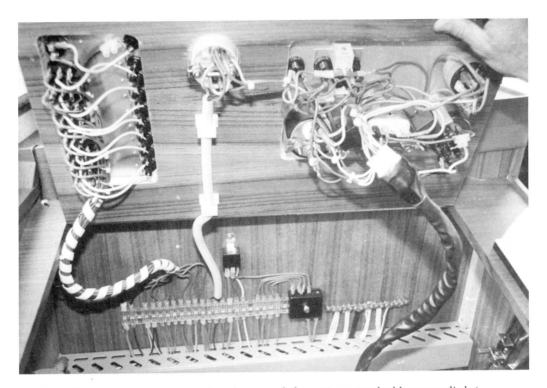

On this dashboard, engine instruments have been coupled up using a standard loom supplied at purchase. A single multi-connector disconnects them. Wiring for other electrical functions is led through one short loom via a push-fit terminal block and thence down into metal conduit for distribution to the various parts of the boat.

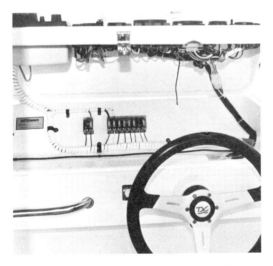

Simple wiring on a small sports boat. An eight-way fuse panel is accessed by lifting the hinged dash.

There are now many ways in which good, permanent connections can be made for cable to cable and cable to equipment. A system of boots and terminations is a very quick and easy method of joining wires and for making connections at gauges, lights and fuses. They can readily be purchased from the local motoring accessory shop and come in kit form with a selection of terminals and insulating boots for a wide variety of cable sizes, together with a special crimping tool – a plier-shaped item with notches at the end which is used to strip cables to the right length, then crimp and secure the terminals. Most of these connectors incorporate 'piggy-back' lugs which allow extra wires to be linked into the system.

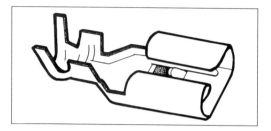

An example of a Lucas Lucar receptacle. The cable is stripped back and 'crimped' into the special clip at the end of the connection.

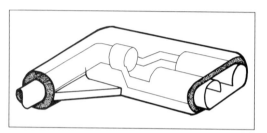

The receptacles can be bought encapsulated in a plastic boot which protects them from damp and accidental short circuits.

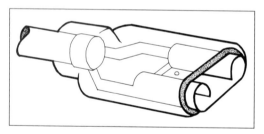

A straight Lucar receptacle with sleeve. These are ideal for connections to switches and instrument gauges.

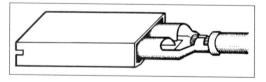

Lucas' Slimlok system comprises sturdy connections that lock together to prevent accidental separation. They can only be separated by pulling on the sleeve, not the cable.

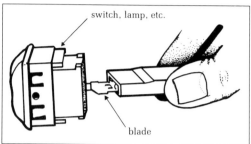

Lucar terminals can be used to make connections to switches, gauges, panel lights and fuse-boxes. They are simple to fit and provide sound, trouble-free couplings.

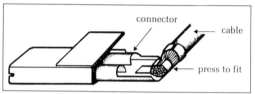

The cable is first clamped into the terminal which is pressed into the sleeve.

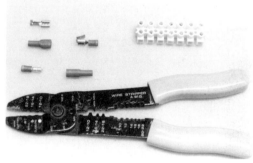

Special crimp cable terminals are available in a range of shapes and sizes and are fitted to the wires using a special 'strip and crimp' tool.

Another method of jointing is to use barrier strip and a sealed box. Barrier strip consists of a series of metal screw terminals that are encased and insulated in plastic with holes at the top to allow access for a screwdriver. They are some-

Small plugs and sockets specially designed for DC use are available for TV or power outlets. They are neat, easy to install and unobtrusive.

times used in the domestic environment and are ideal for use aboard the boat as long as they are encased in a waterproof, sealed box after attachment of the cables.

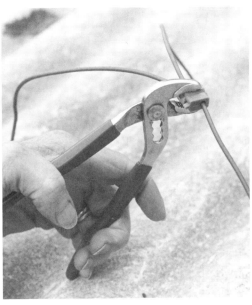

Scotchlok connectors from 3M can be used to splice an extra wire into an existing run without cutting the original cable.

Special die-cast metal boxes with rubber seals around their lids can be bought fairly cheaply from electrical retailers. They are easily drilled, are lightweight and come in a range of sizes.

Care should be taken before making any connection permanent. Strip back only enough insulation to allow a good contact to be made under the screw and twist the strands of the cable together with your fingers or a pair of pliers to neaten the tip.

'Chocolate Block' connectors provide a good, cheap method of cable jointing, but should be encapsulated to prevent corrosion of the terminals.

Barrier strip is a good, economical way of linking wires, but because steel is used for the terminals, the block should be encapsulated in a waterproofed box if used in a marine environment.

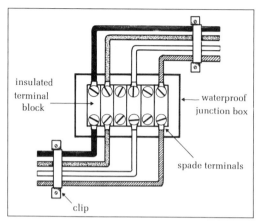

Small, insulated terminal blocks can be used where three or four cables need jointing. The boxes have waterproof lids, sealed with rubber gaskets around the edges.

Loose strands can cause shorting and in any case look unsightly. Before inserting the end of the cable into the terminal, double it back on itself to ensure a snug fit, tightening the screw firmly down on to the conductor.

Never install wiring so that it runs through the bilges of a boat. This is asking for trouble as water will inevitably find its way into the driest bilge on the newest boat and it only takes a small cut in the insulation for disaster to strike. If it is essential for cables to be run through the bilges they should be fed through plastic conduit which can be waterproofed and sealed.

To contradict a popular myth, a soldered joint is not as good as a crimped

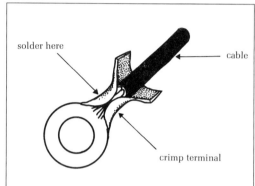

A plate terminal. Once the wire has been crimped in the cable grip it can be soldered to the terminal for extra security.

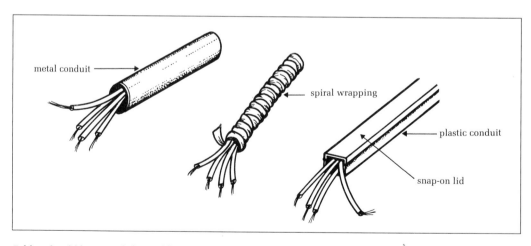

Cables should be routed if possible, in conduits of steel or plastic to protect them and keep them neatly together. In areas where large numbers of wires converge such as instrument panels and distribution boards, plastic spiral loom wrapping material can be used to good effect.

one. A soldered terminal is open to the same sort of vibration that solid conductor cables experience aboard a moving boat. They can fracture causing intermittent open circuits and eventually failure. Soldering should, however, be used to fix certain equipment plugs and sockets such as the co-axial plugs for the VHF radio and Decca navigator. A 12 volt soldering iron, flux and solder should be part of the boat's electrical toolkit along with an iron that can be heated on the gas ring.

Colour Coding of Cables

The two areas where many connections will be required in the system are behind the dashboard and at the main fuse board and distribution panel. Here it is imperative that the wiring is neatly arranged in separate looms, colour coded and tagged for ease of identification. Using special cable ties or wiring loom cord will prevent movement of wires and keep the installation tidy. Where many wires are involved, it might be better to use short lengths of plastic square-section conduit with a snap-on lid. This can be cut to length and screwed down to timber bulkheads before locating the cables. When it comes to faultfinding or adding extra circuits, a nightmare situation can be avoided by having a neat and tidy wiring panel. For maximum legibility, I would advise that wiring colours and tags are entered on to the boat's main wiring diagram, a copy of which should be kept aboard the boat for reference as well as an aid when troubleshooting.

Ready-Made Panels

For the amateur installing from scratch or trying to sort out a spider's web of cabling installed by someone else, the ready-made switch and wiring panels available from electronic shops will provide a neat and practical answer. Some of these panels come complete with a range of switches and even have their own integral wiring loom attached. All the installer needs to do is decide on how many circuits will be required, buy the appropriate panel and couple it up. The amount of switches you will need will depend upon the size and type of boat you own. With a new engine, many manufacturers now supply a ready-made instrument panel complete with

Home-made brackets have been successfully used at the top of this mast to support the masthead light, VHF radio and satellite aerials, with the wiring passing through the mast itself. The use of electrical tape to secure cables is wrong however, and proper cable clips or ties should have been used.

Rubber grommets of various sizes should be used whenever a cable or bunch of wires are required to pass through a bulkhead, metal box or aluminium mast to prevent chafing of the cables.

The owner of this boat has fabricated a simple but sturdy mast upon which he has installed his communications, satellite and Decca navigator antennae, plus navigation light and powerful deck spotlight.

three metres of cable joined by a multi-connector which makes the installation of the engine electrics a much simpler operation.

It is wise to check out periodically the entire electrical fabric for any signs of chafing, corrosion or wear and tear. Look for chafing where cables pass over the edges of timber bulkheads or through holes in cabin walls. Rough edges should be protected by fitting rubber or plastic clip-on edging and by using grommets and glands as specified. Do not attempt to feed too many wires through too small a hole, they can become stretched and will break after a period of time. Switches and contacts should be examined for signs of corrosion caused by dissimilar metals and the action of damp. Clean off contacts and spray with a waterproofing agent such as WD40 or smear with silicone electrical contact grease. When choosing fittings such as switches and lamp holders always go for those that are designated for marine use. Domestic types just will not stand up to the harsh environment present aboard a boat. In the salty environment of a boat corrosion can be rife and most faults in the electrical system can be directly traced

back to it. Once damp, salty air penetrates through to a junction box or switch, a deposit of salt crystal is left behind. Once this dries off it becomes susceptible to atmospheric changes, soaking up moisture and eventually causing short circuits and faults.

Another factor to consider is the type of current used on the boat – usually 12 volt DC. AC rated switches used in the house should never be used on DC currents. The frequency of the alternating 240 volt AC mains is 50Hz which means that the current is being switched on and off rapidly 50 times per second in a positive

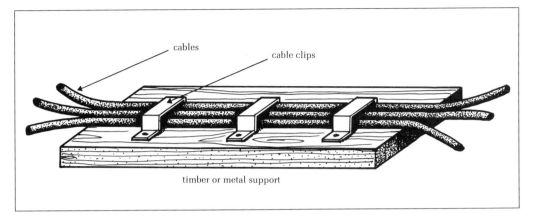

cables

cable clips

timber or metal support

Where cables are required to bridge a gap such as sensor and starter cables coming from an engine block, they should be supported by bridges or tracks to prevent unnecessary movement which could lead to fractures and open circuits.

and negative direction. If you were to switch off an item on an AC circuit; the chances of arcing from the collapse of the current through the opening switch contacts would be minimal. But with a DC current quite a big spark can be produced and therefore switches should be rated to compensate for this. Constant arcing can rapidly burn out an AC switch operating on a DC circuit. This is why it is essential to use only marine-grade materials in the fabric of your electrical system.

SUMMARY

- Take time to plan your electrical system before you start to buy fittings, cable, etc. It will save time and expense later on and will ensure your installation goes smoothly.

- It is very important to ensure that the cables used in any particular circuit are of the correct grade and rating, with enough capacity to carry the currents expected of them.

- Whenever a cable has to pass through a bulkhead, partition or deck it should be protected from chafing by rubber grommets or a special waterproof gland.

- Never use domestic type cable aboard the boat on your DC circuits. The cable will have been rated for AC which flows in a different way to DC. Most domestic cabling is single conductor type which is highly susceptible to the vibrations caused naturally on a boat, leading to eventual fracture, open circuit and failure of the supply.

- The cause of many electrical faults can usually be traced back to poorly made connections at junction boxes, behind switches and gauges. Always ensure that joints and terminals are made using correctly rated junctions and spade, push-fit or ring terminals, clamped and if necessary, soldered to the cable ends.

- Electrical installations can be made simpler and much neater by purchasing ready-made switch and wiring panels. These have all switches and fuses installed in an aluminium or plastic panel which only requires connecting up.

61

6

MAINS POWER AFLOAT

On-board mains electricity of 240 volts was once considered a luxury, but there are so many fixed and portable generators now available that almost every boat owner, no matter what size of craft he owns, should be able to find one for his needs. Mains electricity can be useful in all sorts of ways from powering tools to running the colour television, hair-dryers and even cookers; it is also probably one of the safest forms of on-board heating, cooking and lighting and once correctly installed, is certainly one of the cheapest.

The snag, and there really has to be one

A fixed, engine-driven generator, installed in the engine compartment will provide 240 volt AC for larger boats. On average, and depending upon the equipment to be supplied, around 7KVA will be adequate for most needs.

snag, is one of the initial cost of buying and installing a diesel- or petrol-powered generating set which will be capable of supplying all the needs of a modern boat. For this reason it is usually only the larger and more expensive craft that use electrical power for all of their domestic requirements.

The Fixed Installation Generator

With advances in electrical technology generators have become more compact and can therefore be considered for smaller craft where the size (and noise) of older units would have been a problem. Even relatively high output generating plants – say up to about 15KVA (Kilo Volt/Amps) – are compact enough to be installed in boats of around 35ft and above. For the majority of craft a 6KVA unit would be powerful enough for most domestic requirements. But as already mentioned the cost factor usually puts these units beyond the reach of the average boat owner who probably already has a good gas installation for cooking and perfectly satisfactory 12 volt DC lighting.

For those owners who insist on mains power there is a wide range of sets available, both petrol and diesel driven. Most compact units use alternators which give a greater output from a given physical size and therefore make for a smaller package. Noiseproof covers are usually available to reduce the noise levels acoustically; especially of diesel-driven generators. When they are installed in already sound-proofed engine compartments (the usual place for a fixed installation generator) they reduce noise to an almost inaudible level and, because they utilize the same type of water-cooled components as propulsion engines there are no cooling problems.

For many people the use of on-board mains power would be restricted to the use of power tools when away from a land-based mains supply, and as most DIY boat owners know, an electrical supply makes any job aboard the boat that much quicker and more efficient. It is not until you are without electricity that you appreciate how valuable it is, especially when trying to carry out repairs without it!

For most applications a small, portable four-stroke generator will be ideal. This model supplies 500 watts of AC power – enough to operate small power tools, TV and lighting as well as having a DC take-off for battery charging.

The Portable Generator

Probably the most popular type of generator on the market is the small portable unit which can be carried to where the power is required and is compact enough to be stowed aboard the smallest of boats. The smallest of these usually has an output of around 400 watts and would be powerful enough to run small power tools such as drills, sanders, mini-grinders and soldering irons.

Obviously as the outputs become greater the units become physically larger so that a compromise is often required between power output, size and price.

There are quite a few small portable generators available now and before deciding on a model it is wise to study the individual manufacturers' specification

The AC 'mains' is supplied via a safe plug and socket. The DC 12 volt socket can be seen on the right.

sheets and brochures to decide exactly what your requirements are before purchasing. Some models offer amazing degrees of soundproofing making them extremely quiet in operation but of course these are more expensive and will sometimes have a slightly reduced power output. Some models have little regard to soundproofing but are much cheaper. All the compact portables are petrol driven so due regard must be paid to the carriage and stowage of spare fuel especially if you own a diesel-powered craft and do not normally carry petrol aboard. Many craft have a small dinghy with an outboard motor and already carry petrol aboard.

Engine Output Generators

An alternative to the portable type of generator is to install a mains output generator on the boat's engine. These are also available in a variety of outputs and can be arranged to drive through a clutch arrangement so that the drive is only taken up when power is required. This clutch can be as simple or complex as required. The simplest arrangement is to mount the alternator on a pivoting bracket attached to the engine, driven via a fan belt and pulley. When the generator is not in use the bracket is moved inwards to slacken the belt totally so that no drive is achieved. When power is required the bracket is moved outwards to tighten the belt and is locked into place, perhaps using a simple notched bar.

A more sophisticated system would use an electromagnetic clutch to engage the drive belt thereby allowing remote operation and removing the need to enter the engine compartment to engage the drive as is required by the simpler method. With

this type of generator arrangement it is necessary to set the engine revolutions to a predetermined rate which will enable a steady mains frequency of 50Hz. This can be done roughly by using the engine rev-counter but for a more accurate setting a frequency meter should be installed.

The advantage of using an engine-mounted alternator as a generator is that the power source is already available and soundproofed making the actual installation less arduous. The power required to produce 3KVA, which is enough for most boat owners, will be obtained at little more than tick-over on many inboard engines although a suitable increase of revs ratio will be required between generator and engine pulleys to ensure the correct number of revolutions for the generator to function at a steady rate. This also means that noise levels will be cut to a minimum, as will fuel consumption.

Safety First

With any type of generator producing mains voltage, be it either one of the small portables, a fully installed, plumbed in and silenced large capacity model, or a simple engine-mounted alternator, it MUST be remembered that safety comes first. This means ensuring that safety, both during installation and in use, is a prime consideration. All generator suppliers will be able to give some information on safety precautions, but remember that cocktail of water and electricity, and contact a professional marine electrician for further advice. If you aim for an installation in line with modern domestic requirements but with a greater regard to damp-proofing and cable protection then you should have few problems.

Wiring should be carefully run in conduit (either steel tubing or the plastic type) to protect it from physical damage and engine heat. Proper connections using enclosed junction boxes are another necessity and earth leakage circuit breakers similar in type to the ones you can buy for home use, are another low-cost piece of gear which, when properly installed into the circuit, will add to your peace of mind.

Marina Mains Supplies

Many marinas now have the facility of 240 volt AC mains supplied to each berthing pontoon and plugging into this system when moored allows the boat owner full use of those appliances previously

An example of a plug-in shore electrical point installed at a waterside marina. The electricity is paid for by slot meter.

restricted to the home. For simple jobs around the boat a long extension lead with plug and socket will suffice, but to take full advantage of this type of supply, an AC ring main, similar in type to that installed in the average house can be fitted to the boat. Never use ordinary domestic-type cable for the circuit – special grade rubber insulated cable suitable for use in a marine environment will be required. Two circuits should be fitted; one for lighting and the other for power outlets, and regulations will dictate how many sockets can be fitted on any single ring.

The 240 volt AC system can be considered as being isolated from any other electrical system aboard and utilizes three-core cable for its circuits. The positive, negative and earth wires are returned from the point of entry, the inlet socket on the boat, back to the connection socket on shore. The earth should never be connected to the boat's earth circuit in any way. The circuits should be fused in the following way: a 30 amp circuit breaker or fuse to protect the power circuit ring which is capable of carrying a load of 7kw, and a 5 amp one to protect the lighting ring. Each socket outlet should also be provided with its own fuse rated to the appliance likely to be plugged into it, in addition to the similar protection within the actual plug on the supply lead for that appliance.

Inlet Socket and Wiring

The inlet socket should be mounted in a convenient position on deck which allows ease of access when plugging in the shore connection cable. It should be shrouded from the weather and be of the waterproof type. The male part of the plug/socket assembly should be the end fixed to the

The shore supply cable should be connected to the boat's circuit through a special socket, shrouded from the weather and in a convenient position on deck. Note the waterproof plug at the end of the supply cable.

This socket outlet is protected by a residual current circuit breaker (RCCB). It is rated for use at 240 volt AC at 16 amps.

A 240 volt AC ring can be installed on the boat and connected to a supply on-shore. Outlets are usually domestic-style sockets, but buy the best type you can afford and install them in accessible places, feeding the cable through conduit for a neat and safe job.

boat which means the shore line has a female socket on its end. The reason for this is obvious – allowing the shore lead to be handled in a live state.

Once you have chosen a route for your 240 volt AC ring, select only the best quality socket outlets available. MK are a good brand, and come in switched, fused and non-switched varieties. Always choose the individually switched and fused types, and mount them on special backing boxes which can be screwed directly to bulkheads and the insides of cupboards. If at all possible, choose special marine-grade sockets, but remember that these are likely to be expensive. If using normal domestic sockets waterproof the back connections by filling the patrice box with

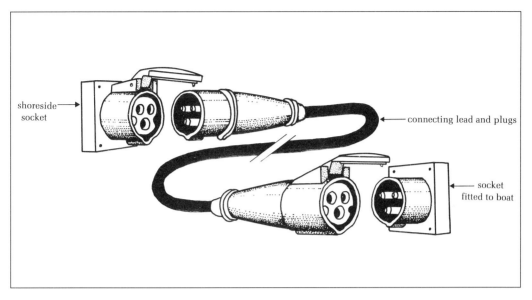

shoreside socket

connecting lead and plugs

socket fitted to boat

Schematic drawing of an AC shore power supply cable. The female socket on the fly lead should be the end used to couple into the boat's plug which will allow the lead to be safely handled even though it is 'live'.

silicone rubber compound and bedding the socket down onto it. Fit a small domestic switched fuse-box with two fuse or circuit breaker outlets, one for the 30 amp power ring, the other for the 5 amp lighting ring and mount it in a similar position to the master fuse panel for the 12 or 24 volt DC circuits, in a dry, waterproof place. For total peace of mind, an isolating transformer may be fitted between this fuse-box and the inlet socket on deck, although most marina supplies should be isolated and fuse-tripped as a matter of course.

SUMMARY

- On many larger craft it will be possible to fit a fixed installation generator to supply 240 volts AC 'mains'. Owners of smaller craft will be able to take advantage of the wide range of portable generators on the market.

- When installing a 240 volt AC ring main on the boat, use only the best quality fittings and ensure that the circuit is correctly fused.

- The fly-lead which connects a shore-based mains supply to a boat should be of the correct grade and size with the female end being plugged into the boat. In this way the cable can be safely handled when in a 'live' state.

7

GALVANIC
CORROSION

Earlier in this book we described how two dissimilar metals when placed in a liquid formed a simple cell or battery. When these two metals were coupled together via a conductor outside the cell, current flowed and one of the metals corroded or dissolved thus providing fuel for the battery.

Let us now apply this situation to our boat. Any boat complete with a metal outboard or inboard engine, coupled to a standard stern tube or propeller shaft is composed of dissimilar metal parts; alloys, brass, zinc, ferrous steel, iron and bronze are just some that may be found associated with the stern gear, engines and supporting brackets that go to make up the average boat's propulsion mechanism.

Now place these metals in the water, add a liberal amount of salt in the case of the sea or some pollution for a river or canal, and the resulting equation is likely to be corrosion in one form or another.

Almost everything aboard a boat corrodes for most of the time that the craft is afloat – unprotected ferrous metals are 'open season' for the ravages of wind and weather, but the type of corrosion most associated with boats is called electrolytic or galvanic corrosion.

Mechanical wear on propellers, shafts and bearings is relatively easy to spot during the course of a service and the remedial repairs required are more often than not indicated by rattling, shuddering and excess vibration under speed. However, electrolytic corrosion can be much

This rudder blade is suffering from severe pitting and flaking caused by galvanic corrosion. The different metals used in the rudder blade, pintals and brackets and the lack of a sacrificial anode have accelerated the problem.

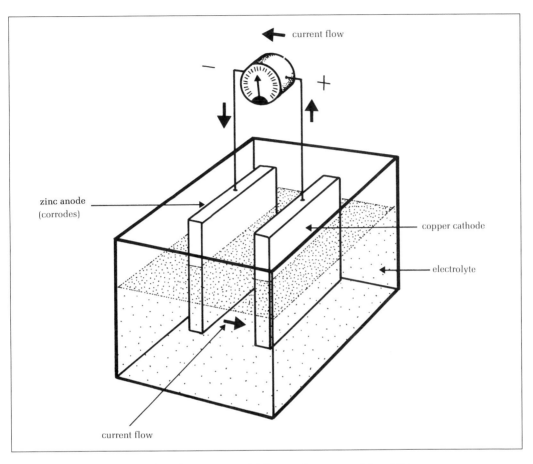

current flow

zinc anode
(corrodes)

copper cathode

electrolyte

current flow

If a metal is immersed in an electrolyte (sea water) it has a voltage potential and if two or more dissimilar metals are immersed and electrically connected outside the electrolyte, a current will flow from the metal with a low potential – called the anode – to the metal with a higher potential – the cathode. When this occurs the anode will begin to corrode. (Courtesy of Lucas Marine.)

harder to locate, especially in its early stages so some care is required when searching out the problem.

Electrolyte

Galvanic corrosion is caused by minute electrical currents flowing between dissimilar metals through the medium of sea water or the badly polluted water of a river. All impure metals, especially alloys of one or more metals such as copper or brass, contain particle pockets which can be labelled as anode and cathode or positive and negative. When these metals are immersed in a conductive medium or electrolyte such as the sea, a primary cell – our battery – is set up causing a small current to flow from the anode or base metal to the cathode or more noble metal.

The result is that the current takes with

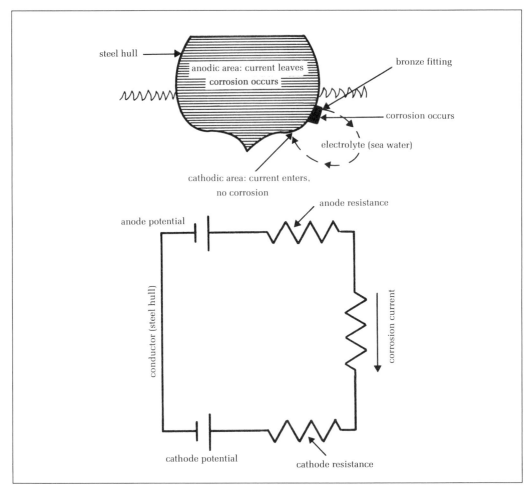

Protection against corrosion can be achieved by the use of sacrificial anodes made of magnesium or zinc. Attached to the hull, rudder blade and/or stern gear they are of a lower potential in comparison to the boat's normal metal fittings. Because of this they act as anodes, forcing the bronze or steel fittings on the boat to become cathodes and corroding away in their place. (Courtesy of Lucas Marine.)

it minute particles of metal from the anode leaving a pitted contusion which is the corrosion that we eventually will see. This is the same process (although more controlled and refined) that is used during the process of electroplating such items as cutlery, car body shells, galvanized sheets of iron and the like.

Galvanic Scale

All metals have a voltage potential and a table can be produced to show the corrosive relationships between most of the more common materials used to build boats, starting with the cathodic or noble (protected) metals and ending with the

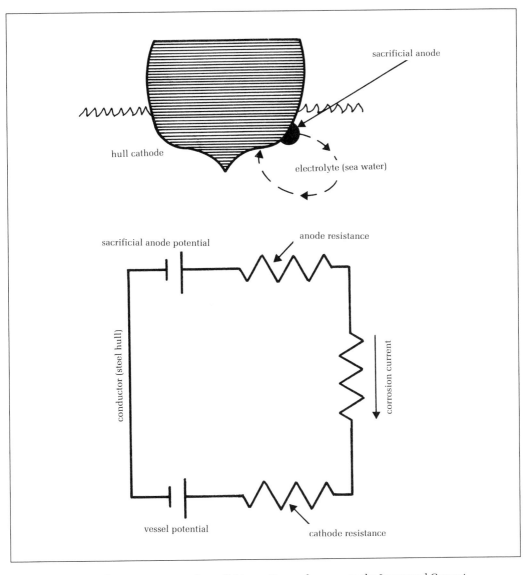

Vessels that have a direct current supply available continuously may use the Impressed Current method of combating galvanic corrosion. Anodes of inert material are placed on the boat and pass sufficient current to ensure that the boat becomes the cathode of the galvanic cell. This method allows a modicum of control depending upon conditions of temperature, salinity or pollution of the water. (Courtesy of Lucas Marine.)

anodic or base metals – those which will readily corrode. As a rule of thumb, it can be stated that whenever a metal has a potential difference greater than 0.25 volts a good deal of corrosion is likely to be the result.

71

The following numerical list shows the voltage potential differences between each of the metals, and from this information it can be calculated which metals are best used with each other for any given situation. It should also be noted that stainless steel is still an active corroder in confined spaces away from the oxygen that protects its oxide skin, but is passive when exposed to the air.

Metals	Voltage
Monel Metal	0.08
Stainless Steel (Passive)	0.08
Silicon Bronze	0.18
Manganese Bronze	0.27
Admiralty Brass	0.29
Copper (and copper paint)	0.36
Stainless Steel (Active)	0.53
Lead	0.55
Cast Iron	0.61
Carbon Steel	0.61
Aluminium	0.75
Zinc	1.05
Galvanized Iron	1.15
Magnesium Alloy	1.60

From the chart it can be seen that metals that are closer together on the scale will have less of a corrosive effect upon each-other than will metals that are farther apart. Surface area too is an important consideration. If the metal forming the anode is smaller than the surrounding cathode, the corrosion will be swifter and much more severe.

Current Leakage

Another type of electrolytic corrosion, probably even more virile than the slow, natural metal to metal variety we have just discussed, is the problem of stray current leakage from the boat's own electrical system. Bad or cracked insulation on wiring will form an easy path for a current leak, as will carbon deposits on switch contacts. Is the wiring from the main supply battery and that of any subsidiary circuits of the correct type and gauge? Cable sizes can be selected by referring to the sizing and capacity tables in Chapter 5.

All boat electrical systems should be provided with a fully insulated earth return system (not a body earth return such as is found on a car). All circuits should be made with two cables, one positive and the other negative. The positive line should be individually fused at the central distribution panel while the negative lead or earth return should be taken back to a common rail or bus bar.

Many boat owners, when wiring up extra instrumentation, or lights, run extra leads direct from the battery; this is bad practice. Extra circuits should be taken directly from spare terminals in the master fuse-box. The battery too, should have a proper isolation switch of the correct size and capacity which will enable the entire electrical system to be turned off when the boat is not in use. This will eliminate any problems of internal current leaks.

Support Your Cables

Leads and cables should be clipped and supported at regular intervals to prevent stretching, breaking of the conductors and eventual failure of the circuit. When making connections, use proper junction boxes or connector blocks and/or crimp terminals and not just 'wrap and tape' joints which will be susceptible to air, damp and corrosion.

Outside the boat, check the navigation

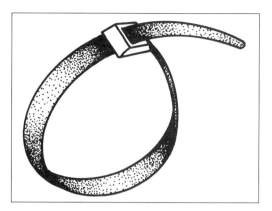

Plastic cable ties are ideal for strapping together large bunches of wires on the main loom. They are cheap, waterproof and simple to use.

lights, deck fittings and cable lead-outs clearing away any corona discharge or green verdigris from around lamp sockets, terminals and switch contacts. Use only waterproof cables of the correct grade on deck and support these in the recommended manner. VHF radio aerials, cables and socket connectors should be regularly greased with electrical lubricant (which can now be bought in a spray can formula from most electrical hobbyist shops) for extra protection and check also the input sockets from shore-based electricity points.

Cavitation Burn

Cavitation burn is very difficult type of corrosion to control, and although it is not directly associated with electrics deserves a mention here. It is caused when minute particles of metal are eaten away from the propeller or its shaft, rudder and associated brackets or even an outdrive leg by the action of cavitation as bubbles of air collapse on the surface of the metal. The

erosion is easily spotted but the cure is not so easily applied.

A badly designed hull and keel near the stern can cause cavitation, as can an outboard motor with an incorrect shaft length. Dirt, barnacles and other projections on the hull can also cause cavitation problems and should be removed as soon as it is practically possible.

Sacrificial Anodes

Protection against galvanic action can be achieved by the judicious use of cathodic protection in the form of self-sacrificing anodes. Usually made from zinc, these blocks are fitted to the exterior of the hull, on the rudder blade, attached to the stern tube or on the 'P' brackets, and are then connected internally to the boat's metal skin, engine or propeller by bonding them via strong, heavy-gauge cabling.

Being very low on the galvanic scale and therefore highly corrosive, the zinc is

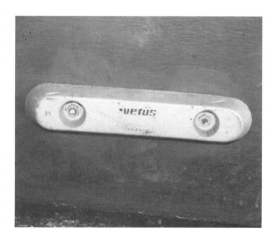

A new anode fitted to the hull of a boat. It will be electrically bonded to internal equipment using cables, bolts and squash washers for good electrical continuity.

73

attacked by the other metals and corrodes away instead of the hull or other fittings. Our simple cell principle comes in again here with the anode becoming one half of the cell. The current flows away from the anode taking particles with it which corrode away into the water, returning via the bonded metal fittings on the boat.

For this system to be totally effective an excellent continuity should exist between the anode and the metal parts to be protected. This is achieved by strapping or bonding the anode blocks to the various fittings using appropriate-sized copper conductors and cables with a substantial cross-section.

For example, a system to protect the stern gear of a boat which might be made up of various metals such as steel, brass, bronze alloys and bearings and is therefore highly susceptible to corrosion, would be to bridge out any flexible couplings, link the bearings and stern tube to the rudder stock (or rudder itself if it is made of metal), the engine and any intermediary brackets with the copper bonding cable, tightly held in place with stud nuts and clamping clips.

Positioning Anodes

The actual positioning of the anode block depends upon the type of boat and its internal engine installation. For larger installations with twin engines and long shafts with two or more support brackets, the anode should be positioned between the bracket and the stern tube and fixed to the boat at a point about midway between the tube and the first bracket.

For shorter shaft installations, the anode

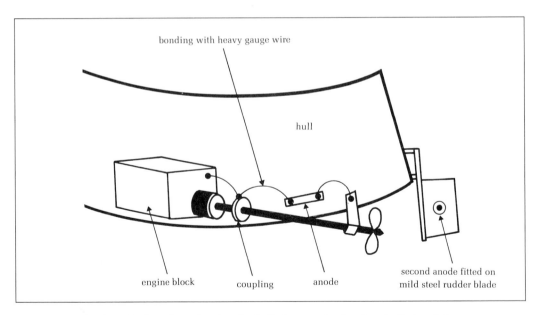

bonding with heavy gauge wire

hull

engine block   coupling   anode   second anode fitted on mild steel rudder blade

Once the sacrificial anodes have been fitted to the hull, they need to be electrically bonded to the equipment they will protect. This is done by using stout conductors of adequate cross-section bolted to the engine block, shaft couplings and 'P' brackets.

can be fixed between the engine and the stern tube. Rudders, stocks and swivels, apart from the stainless steel variety, must have their anode fitted either to the rudder itself or very close to the affected metal parts.

Apart from the engine installation, all sea cocks, inlet valves and water intakes should be bonded and protected by individual anodes. Most suppliers of cathodic protection devices supply anodes and bonding materials in a variety of shapes, sizes and forms. If considering fitting your boat with a set of anodes, you would be well advised to make contact with one of the companies who will be able to provide both practical and technical advice regarding the fitting of a complete system of protection to suit your particular craft be it made of steel, GRP or timber.

Anode Types

There are two main types of anode – zinc and magnesium – and their use will depend very much on the type of waters upon which the boat is regularly cruised. If you do most of your cruising on freshwater such as rivers and canals, the fitting of magnesium anodes should be sufficient to protect your boat. Steel-hulled narrow boats for example, should be fitted with anodes on the hull itself as the material skin is usually no more than a few millimetres in thickness and can not afford to be worn away for very long before extensive and expensive repair work will be needed.

All craft used regularly at sea will require zinc anodic protection in an effort to combat the caustic effect of the salt-

A specially shaped sacrificial anode (bulbous shape) designed to protect the propeller shaft on a yacht.

water electrolyte on its fixtures and fittings. Even glass-fibre boats will require some protection for their stern gear, out-drive legs and outboard engines.

Galvanic corrosion is not only limited to the exterior of the hull. Inside the boat, damp and humid conditions can cause their fair share of corrosion to take place, although this may be on a more localized level. Painting metal surfaces and a thorough pre-paint preparation scheme involving sand-blasting and/or etching of surfaces will help to keep corrosion at bay, but remember that the smallest chip in the paint could be the flaw needed to start up galvanic corrosion.

Once fitted, many owners tend to forget completely about their anodes, but a regular inspection – perhaps even twice per season depending upon the level of pollution in your area – will give a useful monitoring of the state of wear on the anodes. In any event an anode should be

This anode, fitted to the rudder of a motorboat, has been in position for over a year and is now quite wasted and should be replaced if protection is to be maintained.

replaced when it is three-quarters wasted, and as a matter of course after two or three years.

SUMMARY

- Any boat which has different metals used in its construction, especially on underwater equipment like the rudder or propeller shaft, is liable to suffer galvanic corrosion if not properly protected by self-sacrificing zinc or magnesium anodes.

- Once the anodes have been attached to the various parts requiring protection, continuity should be ensured by bonding each item to its neighbour using sturdy cables.

- Inspect your anodes at least once each year, especially if your boat is kept afloat all season. Replace them when they are about three-quarters wasted.

- Stray current leakage from the boat's own electrical system is a virulent form of corrosion. It can be caused by bad or cracked insulation on wiring or carbon deposits on switch contacts. Regular checks and cleaning should keep this problem in check.

- Always remember never to paint over the anodes when repainting a hull. Doing so will render them completely ineffective. If the anodes appear perfectly clean with no corrosion after about two seasons afloat, they are not doing their job and should be checked for bonding and electrical continuity.

8
RADIO INTERFERENCE

Along with the many and varied pieces of electronic gear – some of it very sensitive – available to the boat owner has come the increasing problem of interference from an exterior source. The older the boat the worse this problem can be as little would have been needed in the early days of marine electronics for the suppression or filtering of unwanted Radio Frequency Interference (RFI).

What then is RFI? Basically speaking, it is interference or noise generated in the radio frequency spectrum by a varying or interrupted current flowing in cables, wires or conductors, and is caused by engine ignition circuits, the rotating of the propeller shaft, boat's wiring or even in some extreme cases, radiation from the sea railing, pulpit, etc.

The interference travels from the point of generation to the VHF radio-telephone where it causes the trouble, via the various methods of propagation listed above. The two most important of these are conduction and radiation. The first causes RFI to travel down the boat's cabling or super-structure and can even be extended by utilizing the wiring as an extra radiating aerial.

Cable-Borne Interference

Any cable which carries an alternating current will radiate interference and the bigger the loop or ring formed by the cable the larger the radiated effect will be.

Sometimes, when wiring is bunched together in big loops, the conductors carrying the interference will radiate it to adjacent cables. The boat's VHF radio aerial itself is a prime culprit and a case in point. Interference starts in the cables, travels outwards to the boat's railings and superstructure where it is picked up by the aerial for the VHF, is re-radiated and forms itself into a vicious circle of noise, crackles and unwanted popping which manifests itself over the loudspeaker of the radio.

All this can have a very marked effect on the performance of your VHF radio as well as some of the more sensitive items of electronic navigation equipment such as the digital fluxgate compass, radar or Decca navigation system. How then can the interference be stopped? There are various methods, but the one which you choose will depend to some extent upon the type and source of the generated interference.

Ignition System

The main cause of interference will almost definitely be the petrol engine installation which has always been a prime culprit of ignition-borne RFI. The first checks to make are to ascertain that it is in fact the engine itself rather than a belt-driven

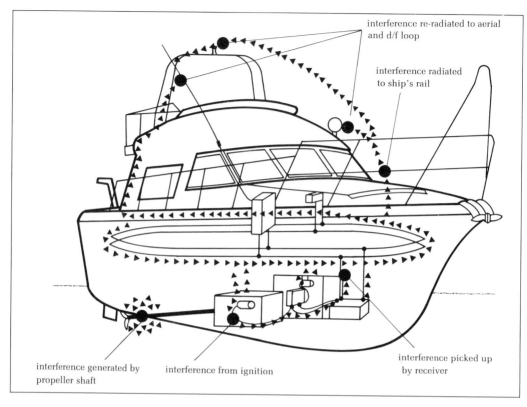

interference re-radiated to aerial and d/f loop

interference radiated to ship's rail

interference picked up by receiver

interference generated by propeller shaft

interference from ignition

Diagram showing the propagation of interference generated from various points on marine craft. (Courtesy of Lucas Marine.)

dynamo or alternator that is causing the problem.

If you stop the engine and take off the field and output leads from the alternator, the noise will persist if it is the engine itself. The level of fizzing will vary with the speed of the engine thus proving the point. The noise can be eliminated or greatly reduced by repositioning the ignition coil. Bolt it directly to the engine block or preferably bond it with stout cable to the block. Fit a suppression capacitor (value 0.5μF to 1μF) between the low tension wire and earth and if all else fails, try shortening the ignition and king lead from the coil, using suppressed leads to each sparking plug. Suppression capaci-

tors can be bought quite easily from a motoring supply shop. They are round in shape and have a single short lead coming from the top, the end of which is fitted with a slip-on tag for attachment to the cable or output lead terminal. The body of the device features a flat slotted clip which allows it to be securely fastened to earth via a nut and bolt on the engine block or equipment housing.

Screened Cables

RFI which is generated by the alternator can be reduced by using screened and sheathed cables and by keeping wiring

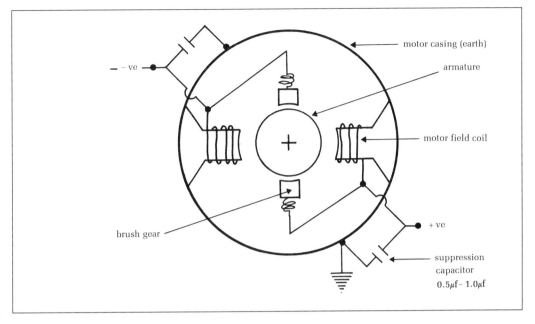

One method of suppressing radio interference caused by motors such as the windscreen wipers or bilge pump, is to fit small capacitors between the field coils of the motor and earth. Most modern equipment will have these fitted during manufacture.

runs as short as possible. The sheathed outers of these cables should also be effectively bonded to common earth. If the alternator is an old one, try replacing the motor brushes and cleaning the commutator arm with an electrical cleansing solvent. As an addition to these precautions a big suppression capacitor can sometimes be used to absorb the AC voltages of 'noise'. This is fitted as a filter into the generator circuitry but must be of a capacity and temperature standard to withstand both the high voltages and heat produced here.

Electric Motors

A noise similar to that of the alternator is

that which is generated by the many types of electric motor on board the boat, for example bilge and domestic water pumps, windscreen wipers and electric winches. These can be successfully suppressed by fitting small, low-value capacitors between the power leads to the individual motor and earth.

This type of motor usually causes a high-pitched buzzing which is constant and present as long as the motor is in operation. Many new motors produced exclusively for use in a marine environment will already have suppression capacitors incorporated within their casings.

Probably the best way of tracing and locating the various sources of shipboard generated RFI is to go through each possible area of noise generation in turn, suppressing and eliminating them as you go. There are two basic methods to be employed in this strategy: screening and bonding.

Screening

Screening normally entails placing a metal box or shield around the offending instrument, thus preventing the noise from radiating outwards and from building itself up into more virulent interference which would be difficult to contain.

These boxes can be made of brass or more cheaply, tin plate. Where cables pass out of the box a small capacitor should be

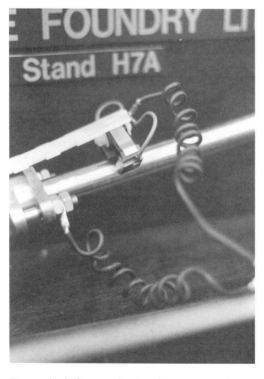

One method of preventing interference created by the propeller shaft is to fit a special 'brush' set. Two graphite brushes run continuously on the shaft and are bonded to earth, suppressing any RFI.

placed between the lead and earth.

As we have said, cables which carry high voltage alternating currents will tend to radiate noise. These cables can be shielded to reduce this effect by covering them in a copper outer (usually a braid similar to that found on television co-axial aerial cable), then encasing the cable in a metal conduit box which performs a similar function to that of the screening boxes mentioned above.

It is very important to ensure a good electrical coupling between the cable, its shield and the conduit box. This is usually achieved by laying earth tape made of thin copper strip beneath the cable and securing the cable at regular intervals along the hull to this tape. If this is not done the copper tape itself could become a radiating aerial and you will have worsened your problem.

Always remember that secured cable on its own is normally insufficient to prevent total radiation and the conduit or earth tape method will usually have to be added as well.

Bonding

An important aspect of any good RFI suppression circuit is the bonding system. You will remember how important this was from Chapter 7 on corrosion, where the self-sacrificing anodes were bonded to the internal metal parts of the boat by stout cables having a substantial cross-section.

The bonding circuit provides the 'ground' or earth return for all equipment used aboard the boat and will also take fixtures and fittings (pipework, rails, etc.) to the same electrical potential. Normally this circuit ends at a cathodic plate situated below the water-line, and wherever possible instruments should be bonded or clamped directly to the bonding line. Should this prove impractical it should be done with thick cable.

On some of the smaller glass-fibre cruisers it may be possible to create a suitable earthing point by spreading tin foil inside the hull below the water-line. Termination of the earth frame should be to the nearest bulkhead and should be done with a nut and bolt.

If you discover that it is your propeller shaft that is causing the interference, special bonding brushes which still allow the shaft to turn, yet will provide a good earthing strap, can be obtained from certain marine electrical manufacturers.

Other Sources of RFI

Other items of equipment aboard the boat which may be sources of radiated noise include the radar system. This, if badly installed, can be quite an intense source of radiated noise due to the very high pulses of energy employed in its operation. The rapid climb and fall rate of this energy may be a particular problem on the smaller boat when it may not just be a case of radiated noise, but also the radar equipment being scrambled by excessive amounts of radiation emanating from VHF radios in close proximity to it.

Radar scanner units should be very carefully sited well away from the cables and aerials of the boat's VHF and, if necessary, these cables must themselves be re-sited. In normal circumstances a distance of 2–2.5m (6–8ft) should be left between the aerial and the radar scanner.

The subject of Radio Frequency Interference is a very complex one and we have only scratched the surface here. Anyone who suffers from it should really consult the manufacturers of the equipment affected, although most modern marine electronic gear is very well suppressed anyway. If the manufacturer can not help you, and you are still suffering after trying the simple suppression cures listed in this book, then a marine electrical company specializing in the subject, such as Lucas Marine, would be the people to contact.

Remember, if your boat's radio communication equipment is being acted upon by noise in the distress frequency bands (500–2182kHz) or the Radio Direction Finder (RDF) wavebands (200–400kHz) then the source of this noise *must* be found and stopped.

SUMMARY

- Any cable carrying current will radiate some form of interference, and the bigger the loop of cable, the larger the radiated effect.

- Interference from electric motors and engine ignition systems can usually be suppressed by fitting special capacitors available from most good motoring accessory shops.

- The VHF radio can be acted upon by interference and should be mounted well away from potential sources of 'noise' such as windscreen wipers, the radar and fluorescent lights.

9
THE VHF RADIO

Probably the most vital piece of electrical equipment aboard the boat is the VHF communications radio. In the event of an emergency occurring while at sea, it is by far and away the best method of summoning help quickly and efficiently. It takes but a few seconds to switch to Channel 16 (the Call and Distress channel) and transmit your message – much faster than setting off flares – and with the comforting knowledge that your urgent call will have been heard by a sympathetic ear.

The VHF is also a useful link to the land-based telephone network and can be used to glean a wide variety of information on weather forecasts, navigational warnings, access to the coastguard or even to book your overnight berth at a distant marina.

Radio lighthouse beacons can be used with some VHF equipment to obtain bearings and several coastguard stations use VHF direction-finding apparatus, operating on the VHF frequencies, which enables them to give you your position if you are fog-bound or if you have come

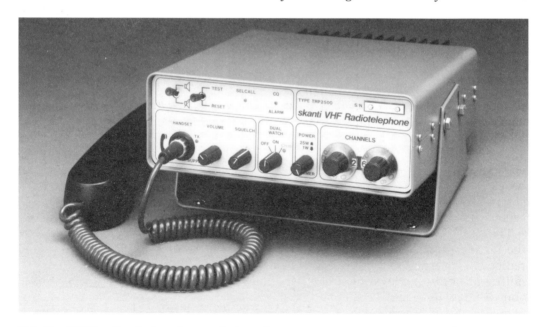

This Skanti VHF radio-telephone features dual watch, two power output settings, a range of channels and a phone-type handset.

unstuck with your navigation and have temporarily lost your way.

So, what in fact does the abbreviation VHF stand for? It stands for Very High Frequency and is used as a blanket term to include all frequencies in the group 30–300 MHz, with wavelengths from 1–10 metres. It is mainly used for communication between fire-fighters, the police, army, RAF and navy, as well as for local and national radio broadcasting and, of course, marine use. The section of the VHF frequency spectrum that has been allocated to marine usage is 156.025 MHz to 162.025 MHz. This frequency band is then further divided into separate channels, each 25 kHz wide, which are numbered and given a specific function.

Simplex and Duplex

The 25 kHz channels comprise two separate types, namely Simplex and Duplex. The Simplex channels utilize a single frequency, and the radio user transmits and receives calls on this one frequency. With a Duplex system, as the name implies, two frequencies are employed, one is used to transmit and the other to receive. Most modern VHF sets are tuned to operate on the Duplex system, enabling normal two-way conversations to be carried out without the need to be continuously saying 'over' at the end of each sentence.

There are about 57 separate channels in the marine VHF band and many modern radios are equipped to receive and transmit on all of them. However, in practice you will only use about 12 of them. Some of the less expensive radios and hand-held types only incorporate this number of channels and some even have the facility

to allow you to switch to alternative channels applicable to a particular cruising area.

Range and Aerials

The range of antennae or aerials available for the VHF user is quite large with aerial sizes and types to suit all boats and prices. The general range of transmission on VHF is short, depending upon the power output of the set, the atmospheric conditions prevailing at the time and the location of the vessel. It is judged to be approximately 30–35 miles but this figure can vary dramatically depending upon the afore-

A variety of VHF radio aerials are available for different boats. These whip aerials are all supplied with a length of special co-axial cable which should not be shortened on installation.

mentioned factors. Sometimes the distance of transmit/receive can be extended to hundreds of miles, while at other times freak atmospheric conditions may cause almost total suppression of the signal as soon as it leaves the aerial.

Today, all VHF radios are solid state devices, using entirely passive components, integrated circuits and transistors. They are usually very reliable and are simple to install. It is, however, vital to install the set's aerial correctly as a poor installation can have a dramatic effect on the performance of the radio itself. The power output of a VHF radio is limited by law to a maximum of 25 watts, so the best use of this available 'umph' is paramount to the radio's performance at sea. As the aerial is the last link in the chain of events that leads to efficient transmission and receiving, the importance of its siting and matching to the set is obvious.

Basically, the higher the aerial is on the boat the better. Use a good quality make and a length of proper co-axial cable to couple it up to the radio. Ensure that couplings and plugs are well fitted and when connecting, take care not to short out the aerial input with the radio switched on as this could seriously damage the output stage and might even invalidate your guarantee.

Regulations for Use

It should be remembered that by its very nature the marine VHF band is a crowded medium and the rules that do exist have been created to ensure minimum congestion and maximum legibility.

There are three main groups of frequencies allocated for special functions; these are inter-ship communication, public correspondence (for talking to coastal stations and for link calls to the British Telecom network) and port operations which include harbour authorities and tug facilities.

The most important channel with priority over all others is the Call and Distress channel – Channel 16 – and an understanding of its use and operation is essential. The Call and Distress channel was established to eliminate to a great extent the confusion and chaos that would arise in an emergency from a vessel trying to contact other vessels who may be on any one of the other 57 frequencies available. Imagine trying to summon help and having to switch from one channel to another repeating your Mayday message over and over again in the vain hope that someone is listening.

To prevent this, Channel 16 was introduced, and designated a Listening or Call and Distress channel. All main coastal stations monitor it 24 hours a day, 365 days a year, and all ships also keep a listening watch. So, to call another vessel, a coastal station, the coastguard or the marina it is only necessary to monitor and use Channel 16. Once contact has been established, both parties then select another working channel by mutual agreement. Normal conversation should never be carried out on Channel 16.

Because of the widespread monitoring of Channel 16, it is obvious that this is the best channel on which to broadcast your call for help, and distress calls should always be made on it using the correct procedures and terminology which are fully covered later in this chapter. The Royal Yachting Association booklet G22 lists the correct methods for the use of this frequency and the variations for different types of vessels.

Mayday and PAN-PAN

A call on Channel 16 and the use of the word 'Mayday' indicates that a vessel is in grave danger and requires immediate help. The Mayday procedure is simple:

'Mayday, Mayday, Mayday. This is *Royal Dart, Royal Dart, Royal Dart.* MAYDAY *Royal Dart.* My position is (Give latitude and longitude or true bearing and distance from a charted point, and the nature of distress – on fire, aground, etc). Over.'

This call takes priority over all other traffic and imposes a general radio silence. If there is no response, check that the radio is switched on, that power is connected and that the aerial is correctly rigged – it sounds stupid, but it has been known! – then repeat the call. If a ship or land base is listening you can expect an immediate answer. If not, repeat several times and then, if necessary, try another channel. Any other channel may be used to transmit a Mayday call.

The words PAN-PAN are used for lesser emergencies where a Mayday call is not necessary and is also broadcast on Channel 16. It is usually employed when a vessel needs to make an Urgency Call regarding the safety of its crew or the ship itself. An Urgency Signal is broadcast in the same way as the Mayday procedure except that the words PAN-PAN are substituted for Mayday.

Securité

Securité is the word that is used mainly by coastal stations to herald the broadcast of a general navigational warning; a gale warning, a buoy that may have dragged its station, an extinguished beacon or a vessel adrift. The initial announcement is made on Channel 16 and an instruction is given to the skipper to change to the channel selected by the coastal station. When this has been done, the full message and information will be broadcast.

Dual Watch Facility

Sometimes it may be prudent to monitor both Channel 16 and the local marina or port frequency, and some VHF radios incorporate a dual watch facility comprising special circuitry that automatically switches between the two channels. When a transmission comes through on one or other of the monitored frequencies, the radio locks on to that particular channel, whilst still retaining priority for Channel 16. This eliminates the need to monitor and determine which channel is carrying the message.

Once you have bought a VHF radio you have to be licensed to use it. The documents required are a Ships Licence and a Certificate of Competence in Radiotelephony or Operators Licence. To get your certificate means sitting an uncomplicated exam which is usually monitored by the RYA and takes about one hour. It consists of a written paper of some fifteen questions which is followed by a practical operator's test on a VHF simulator. The full syllabus and a selection of test questions is available in a special booklet from the RYA called G26. Once the examination has been passed, the user is entitled to operate a VHF ships radiotelephone in any British ship in any part of the world for the rest of his life. It is illegal not to hold such a certificate.

The radio itself must also be licensed. An application form will usually be supplied by the dealer who sold you the set but one can be obtained from the Department of Trade and Industry. It is renewable on an annual basis and must be kept on board the boat for inspection if necessary. If you can not produce your licence, you might find your set being confiscated.

Fitting the Radio

Many radios are about the same size and most are supplied for use on either 12 or 24 volt DC systems. A standard bracket is the usual method for fixing the set to the boat and this will allow the unit to be mounted in a deck-head, overhead at the wheelhouse, on a dash-type console or even just surface mounted. Usually all the controls are located along with the speaker on the front of the set, with some exceptions where controls can be on a remote unit,

VHF radios can be mounted in a variety of positions to suit a particular boat lay-out. This bracket allows the radio to be mounted on a dashboard or overhead, above the helm position.

For open boats, the VHF radio may be protected by fitting a waterproof cover and box. The controls are duplicated on the front of the cover and are attached to the main radio controls inside.

near the helm say, or the speaker can be incorporated along with the microphone in a telephone-type handset.

Although most VHFs will have been designed specifically for use in a wet environment, care should still be exercised when siting the unit on board the boat. It should be placed in a convenient position not only for ease of use, but also for splicing into the power supply and routing the aerial cable and finally, should be out of any direct spray from the sea. Small plastic covers can be bought to fit over the front escutcheon plate on popular models which afford good protection from spray. This is especially important in small, open sports boats.

It is also important to allow for a free flow of air to circulate around the back of the set. Some of the more powerful types have big output transistors bolted to the back which need to be kept cool. A small

vent should be fitted near the radio in order to give the required ventilation.

Remember also to site the set well away from other electronic devices and especially the boat's compass which could be seriously decalibrated by the sheer mass of metal or by the magnet in the speaker. At least 0.5m of separation should suffice.

There are usually two connections to be made when fitting a VHF. One is the 12 or 24 volt DC power supply cable and the other is the co-axial aerial cable from the main transmit/receive aerial. The power supply cables are sometimes supplied with the set and will therefore be of sufficient thickness to carry the power needed to work it. The set must be separately fused and this is usually done either internally or with an in-line fuse-holder. A 6 or 7 amp quick-blow fuse should be sufficient in most cases.

If you find that you need to extend the

The connections for most radios are simple and comprise the power connection and aerial socket.

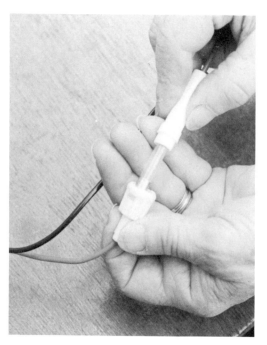

Always protect the radio power supply by installing an in-line fuse and holder in the positive feed. Manufacturer's instructions will give the correct fuse rating for any particular radio.

speaker cables, make sure that you use the same or a heavier gauge of wire. About 10 or 12 gauge will do, but in any case it is best to keep cable runs as short as possible and also make them accessible for inspection. Run them in special conduit if at all practicable. If the cable needs to pass through a deck bulkhead or other barrier, a rubber grommet should be used to prevent chafing. Avoid if possible running the cables near other high power wiring as possible RF interference to your transmissions/reception may result. If cables need to pass outside the craft, use a waterproof deck gland. These are available in many sizes from your local chandler.

Siting the Aerial

The VHF aerial is a very important part of your installation. It is the final link in the chain that forms your transmit/receive line and should be the best that you can afford. Aerials do not usually come supplied when you buy a VHF set, but there are many types to choose from and you are bound to find one that is matched both to your set and pocket.

Aerials range from the big whip types for the larger craft through to short helical ones with the aerial wound round a short tube of rubber. Whichever type you fit, ensure that it is mounted as high up on the boat as possible for the best reception. A small type can be good on a high-sided craft and are easy to remove as a security measure, but on the smaller sports boat,

Many VHF radio aerials are supplied with a mounting bracket that allows the aerial to be raked at any angle to suit the boat. It is also essential to be able to lower the aerial if passing under low bridges.

lying lower in the water, one of the lightweight GRP whip aerials may be best.

They are fitted by attaching them to a small stainless steel bracket, to a mast or the deck via a waterproof gasket. The angle of the aerial can be adjusted to suit reception and the lines of the craft. Most aerials now available will work on the frequency range 156MHz to 162MHz and all are supplied with a good length of co-axial cable which should be long enough to pass through the boat from aerial to radio without needing a joint. If you do have to make a join, ensure that you use proper co-axial couplers as signal loss could be the result of a poor, twisted or even soldered joint wrapped around with tape.

Suppression

Finally a word about interference suppression. VHF radios operate in the 150MHz frequency band and are usually quite free from RF interference. However, sometimes certain electrical systems aboard the boat can cause problems. These include the ignition circuits of engines, alternators, windscreen wipers and other on-board motors. Certain voltage regulators can cause trouble if they are of the older type, so be careful once again about the siting of your radio set.

If you get really serious problems with interference you can buy special capacitors and coils which can be fitted to the armatures of motors and across the motor windings. These are available in a range of values from your local car radio accessory shop which will also be able to give advice on their correct fitting. For further information on interference suppression, *see* Chapter 8.

The installation of a VHF radio is really a simple operation which can be an enjoyable job to do over the winter lay-up period. If you do get stuck, most manufacturers will be able to help or, as a last resort, you could think about calling in a qualified marine electrician.

SUMMARY

- Channel 16 is the Call and Distress channel and is monitored constantly by shore-based organizations such as the Coastguard. A listening watch is also kept by all ships who should have their radios set to monitor Channel 16.

- Once you have bought your radio you will need to have it licensed by applying to the DoT. You will also need a certificate of competence which can be obtained by sitting a simple RYA examination.

- The aerial and its position on the boat is most important in order to get the best performance from the set. It should be fitted as high as possible. Remember not to shorten the cable as it will be 'tuned' to the individual set.

- Care should be exercised when siting the VHF radio to avoid areas which will get wet as the boat is used. This is especially important on small, open sports boats where spray regularly comes aboard. Protective plastic covers can be bought that enclose the front panel of the radio but which still allow the controls to be manipulated.

10
THE ECHO SOUNDER

Next to the VHF radio-telephone, the item of electronic equipment that is used probably more than any other aboard the boat is the echo sounder. This is an essential tool for anyone who regularly cruises in shoal or coastal waters as it paints an accurate and in many cases, comprehensive picture of just what is going on beneath the hull of the boat. As a navigator you should be just as concerned about the depth of water and state of the ground below your hull as the actual position of the boat in the water. The echo or depth sounder gives you this information. It is also one of the simplest and oldest pieces of all the electronic navigation equipment aboard and it is now quite rare to find a boat without one. Costs have come down considerably with the production line methods now employed by electronics companies and with the advent of microprocessor technology, and this has resulted in not only reasonable prices but also a wide and diverse range of sounders available.

The most basic of types can cost as little as £60 and if used correctly can be just as much use aboard the boat as models costing several hundred pounds. The choice will depend on what you will require from your sounder, the information and facilities it offers and the depth of your pocket.

How they Work

The basic principle of operation is similar for all types of sounder. A transducer below the hull of the boat generates a pulse of energy, usually in the range of 150kHz, which is sent down through the water. This pulse strikes the seabed and is reflected back where it is received by the same transducer. The electronics in the head unit at the helm unscrambles the information and measures accurately the time taken between the pulse transmission and reception. The nominal speed of sound in water is approximately 4,730 feet per second and knowing this fact the computer in the sounder can calculate the distance travelled by the sound pulses – half this distance will be the depth between the transducer and the seabed.

One of the main indicators to the quality of a particular sounder will be the pulse generating crystal in the transducer. If the crystal is of poor quality the readings will not be accurate. Another factor is the method the sounder uses to display the information gathered. It could be a simple rotating series of Light Emitting Diodes (LEDs), a Liquid Crystal screen made up of thousands of pixels with full colour showing changes in temperature and different types of ground, or a paper print-out on a roll. All these methods have their advantages, but it comes down to individual requirements and personal preference.

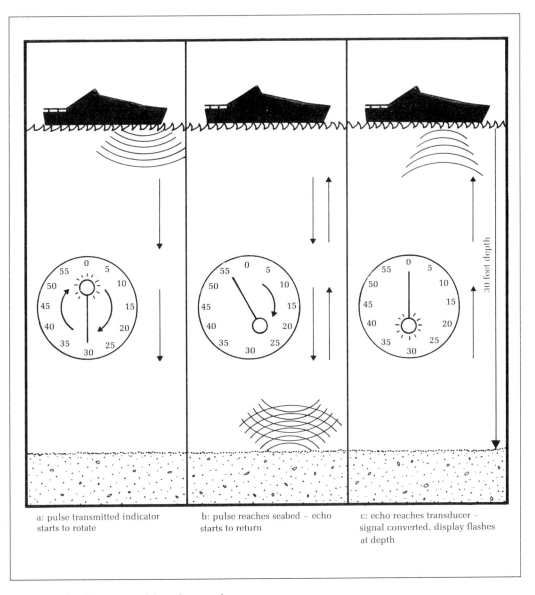

a: pulse transmitted indicator
starts to rotate

b: pulse reaches seabed – echo
starts to return

c: echo reaches transducer –
signal converted, display flashes
at depth

The principle of operation of the echo sounder.

Most of the echo sounders on the market opt for a transducer that assumes 4,730ft/sec which tends to show a slightly lower depth than is actually there, giving a useful margin of safety. More sophisticated sounders offer correction and trim facilities for salt water, pollution and temperature, although in practice these factors would not concern the average boat owner.

Performance

When it comes to choosing a sounder the transmission frequency could also have an important bearing on its overall performance. Lower frequencies down to around 30kHz give a better penetration of the water and thus allow the sounder to operate in deeper waters. For most applications though, the 150kHz frequency will be more than adequate.

The width of the beam splaying out from the transducer is another important factor to take into consideration. A wide beam width is good for a wide sweep and will allow the sounder to carry on giving readings even in rough seas when the boat is pitching and tossing. It does, however, reduce the power of the pulse signal and might even give poor readings in heavy saline or badly polluted zones. A narrower beam width gives excellent readings in deep water and boats that regularly cruise the oceans will normally be fitted with this type. The angle of the beam as it leaves the transducer is about 15 degrees whereas sounders used mainly in shallower waters have a beam angle of about 50 degrees. Most of the sounders that can be bought for the pleasure boat in the UK use 50 degree beam transducers, which are smaller than the narrow beam types and are much more simple to install.

The rate at which the pulses are transmitted is another variable that must be taken into consideration. Sounders used on small boats tend to have a pulse rate of between 4 and 25 pulses per second and this rate varies according to the depth at which the sounder is operating. The shallower the depth, the higher the pulse rate is the usual equation. Each sounder will have a minimum depth at which it can operate successfully and this is controlled by the pulse rate. The transducer cannot simultaneously transmit and receive return pulses, which means that you are unlikely to get very accurate readings at depths of less than about a metre.

Display Types

As we have already said, there are several ways in which the sounder 'head' can present the depth information to the end user. One of the earliest types was the flasher, which originally used a neon light but then moved on to LEDs. These are usually found on the cheapest sounders such as those that used to be produced by Seafarer. They have been around a long time, are easy to read and rarely go wrong. Many buyers of second-hand craft will find one of these units fitted and most will still be working away giving an efficient and reliable service. Some of the earlier models only had one scale of readings,

The simplest and oldest form of echo sounder, this model has a display made up of light emitting diodes mounted on a rotating disc. It is both accurate and simple to use.

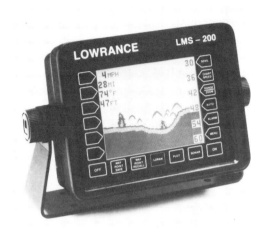

The liquid crystal echo sounder. The display is made up of several thousand pixels and the electronics process information such as water temperature, boat speed and depth in feet or metres, as well as showing a graphic representation of the seabed.

usually measured in fathoms, but these later graduated to a switchable scale to provide both feet and fathoms with the depth in metres as a third option. The modern flashing sounder now features only two scales – metres and feet.

Digital-only models are now very popular for their ease of reading and relatively small size – usually being contained in smaller flush-fit panels which look good at helm or flybridge. Another advantage is that the digital sounder is fitted with a memory that is capable of storing approximately ten consecutive depth readings, analysing them and giving an average read-out on the display. They are much easier for the high-speed sports boat user than the flashing type because their large figure displays are much easier to read when the boat is

A paper roll sounder. The paper moves slowly and a pen inscribes the resulting echoes and depth information. This type is useful if you want to keep a permanent record of your soundings, for example when wreck diving or deep sea fishing.

bouncing off the tops of waves. At night, the displays are backlit by 'quiet' red or green lamps.

The analogue unit is probably the least common of all the types mentioned and these are mainly used as repeaters or second displays to a main unit. They are usually quite simple in their operation and are also easy to read, the display being in the form of a gauge with a needle pointing the depth. Older yachts may still have this type fitted and they can be useful as a back-up on power cruisers or yachts if the electrical supply to the main echo sounder fails.

The paper-type sounder is usually much more bulky than any of the models mentioned, because it needs to be able to accommodate the mechanics for the paper roll and its transport. This type was usually referred to as a 'fish-finder' and is still mainly used by fishing fleets today. It could also be useful for yachtsmen who might want to keep a permanent record of

A video echo sounder is more bulky than the LED type, but capable of displaying more detailed information on its cathode ray tube.

their soundings and surveyors and deep sea divers could benefit when underwater map making or wreck finding. The paper sounder gives a permanent record of the seabed and depths by utilizing a pen trace transcribing the information onto a roll of paper using the data collected from the transducer. The system shows all the information recorded by the flasher sounder but in far greater detail and in a form that will produce a very clear picture of what is happening below the hull of the boat. For wreck finders the rolls of paper records can be marked accurately with the positions of sunken boats, coral reefs or any other underwater item of interest and may be linked with the appropriate navigation chart of the area enabling you to build up a complete picture of the seabed and that which lies upon it.

Since the advent of micro-electronics, the video screen echo sounder has been developed and could be considered the state of the art. It does a very similar job to that of the paper type and some units have a memory store which allows you to replay a certain segment of your soundings for referral. It does not, however, provide a permanent record in the same way as a paper sounder. The development of this type of sounder has taken the display from a grainy, smudged black and white display to a sharp, crystal-clear, multi-coloured one using several tones to distinguish different depth levels and even different types of silt or rock on the seabed. They are much more expensive than their simpler brothers, but offer more in displayed information and ease of use.

Fitting the Transducer

However much money you pay for your

echo sounder it will be wasted if the 'business' end – the transducer – is not installed correctly. Poorly sited transducers and those which have been fitted at an angle can play havoc with the accuracy of the final readings. On a small GRP boat it is best to install the transducer on the inside of the hull. This is easier to do as the boat does not need to be taken out of the water and it also makes servicing or replacement simple. In fact, in most small boats it would be better to fit an in-hull transducer, the only exception being steel-hulled craft which will suffer from beam deflection.

The best way to install a transducer in-hull is to mount it vertically in a special rig at the lowest point of the boat (usually the bilges) in a small bath of oil. This oil, which can be bought specially from echo sounder manufacturers or good chandlers, must be in contact with the hull and the transmitting face of the transducer at all times, including those times when the boat is keeled over or rocking through a rough sea. This factor is especially important aboard sailing yachts which might heel over considerably at any given time.

Once the transducer has been fitted into its oil bath the whole item needs to be bonded securely to the hull to prevent any leakage of oil. The transducer face should not touch the hull itself but should be mounted a millimetre or so back from the GRP. Regular checks should be made to ensure that no oil is leaking from the reservoir otherwise in time, the lack of oil will result in no reading as the sounder tries to send pulses through thin air!

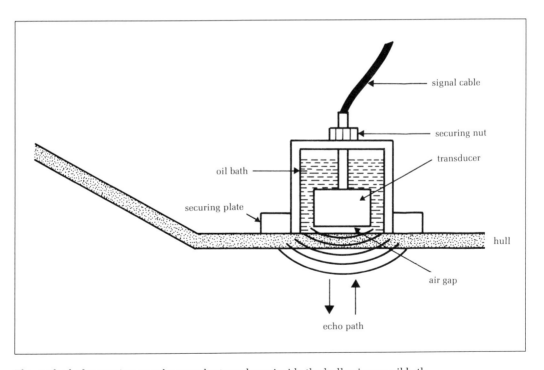

The method of mounting an echo sounder transducer inside the hull using an oil bath.

The component parts of an echo sounder transducer – (left to right) securing bolts, transducer mounted in fairing block, rubber sealing gasket and backing plate.

Through-Hull Transducers

When fitting a through-hull transducer, the first consideration should be the selection of the correct position on the hull. The transducer should be kept clear of areas of turbulence such as near the propeller or underwater obstructions such as bilge keels, and should also be far enough away from the keel to prevent a cutting off of the beam spread which will reduce the accuracy and range of the unit. The lowest possible position on the hull should be chosen and the transducer mounted so that the face points vertically down.

Obviously the job can really only be done with the boat out of the water, so slip her and thoroughly clean off any algae, barnacles and accumulated dirt from the chosen installation area. To protect the head of the transducer and to assist with the smooth flow of water across its face, make up a small fairing block which can be fabricated from hardwood or glass fibre. This block must be shaped to fit the vee of the hull at the point where you site

your transducer. The block is fitted to the hull by bedding it down on mastic and using stainless steel bolts through to backing plates on the inside. Never install a transducer in a fairing block so that it is an immovably tight fit. When working, the transducer vibrates or oscillates and this could be prevented if the unit is too tight.

A matching block is then fitted to the inside of the hull with the back end of the long transducer thread passing through it. The whole assembly is then secured by tightening the single backing nut and washers on the transducer stem.

Paint both blocks for maximum protection. The outer one can be given a good coat of epoxy light primer which is a paint suitable for use on wood, steel or GRP boats. The inside one can be painted using a metallic pink primer. Remember to bolt the two blocks together using plenty of mastic, after which the cable from the transducer can be passed through the hull from the outside. Use further sealant on the inside before the backing nut is tightened.

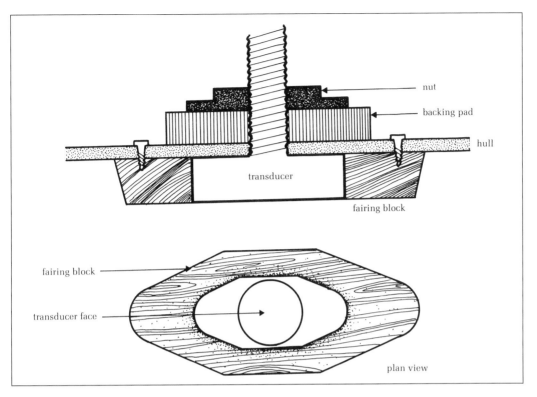

nut

backing pad

hull

transducer

fairing block

fairing block

transducer face

plan view

Cross-section of hull showing external echo sounder transducer. The fairing block protects the transducer from damage and keeps turbulence to a minimum.

The transducer in position looking from the inside of the hull with connection cable coiled on the left.

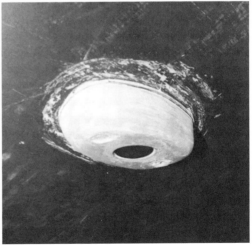

Fairing block sealed in position looking from the outside of the hull. The transducer face can be clearly seen.

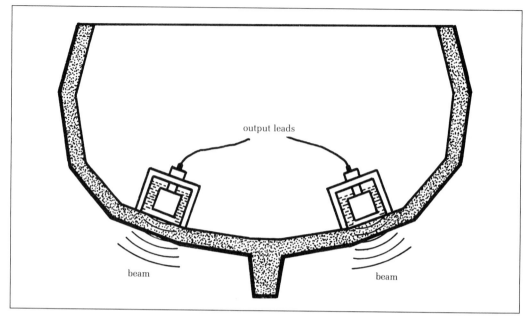

output leads

beam

beam

Twin echo sounder transducers are often fitted into yachts that tend to heel over underway.
Overlapping the beam angles ensures that at least one beam will point vertically downwards.

Once this task has been completed there is little more to do here other than to touch up your bottom paintwork, including a very light coat of anti-fouling over the face of the echo sounder transducer, before moving inside to complete the main unit installation.

Fitting the Sounder Head

Installing the echo sounder unit should provide little in the way of problems. Remember to route the cables from the transducer away from sharp corners and if a cable needs to pass through a bulkhead, ensure that the hole is fitted with a protective rubber grommet. Route the cables in the shortest and most convenient way possible and clip them up at regular intervals to prevent loops of wire being caught up on items of ship's gear. If possible, run the entire length of cable in some plastic conduit. This can be cheaply bought from DIY shops and will prevent your cable from becoming snagged or ripped. An exposed cable that has had its protective outer sleeving damaged will soon act like a wick, soaking up water and rendering the sounder useless in a very short period.

Never cut the transducer cable short, even if you have metres of it left over at the end of your installation. The cable length will have been carefully calculated and 'tuned' to the echo sounder electronics. Shortening it could affect the performance and accuracy of your set. Any left-over cable should be carefully coiled up, tied and clipped safely out of the way.

If the transducer itself needs replacement, this can be simply done on in-hull types by opening the oil bath and removing the unit. Remember though, to refill with enough oil after plugging in the new unit. It is a good idea to make some sort of cover to protect the protruding part of the transducer, its plug and cable from accidental damage. An old plastic oil drum or polyurethane bucket placed over the unit should suffice.

Mounting the reading head should be done after working out the best possible position at the helm. Some boats have their sounders fitted near the chart table but I think that by far the best position is near the helm. Here the helmsman can keep a watchful eye on the display when manoeuvering in and out of potentially tricky situations such as a strange harbour at night. If the boat owner regularly cruises single-handed it is essential to mount the device where it can be seen from the helm.

Many boats have enough space to install the display unit with a flush panel readout that matches the other instruments on board. Certain manufacturers also make other instruments; log, Decca navigator, speedometer, trim tabs, etc., with matching bezels that look both smart and are functional – likewise for the flybridge repeater unit which is usually smaller than the main one down below.

These repeaters are becoming more important as the number of craft fitted with flybridges increases. It is usual with this type of craft to control mooring, inshore cruising and anchoring from the flybridge, and it is therefore equally important for the helmsman to know the depth of water beneath his vessel from this position.

For smaller cruisers the sounder can be mounted on a multi-angled bracket either above or on the dash itself. This bracket can also be used to mount the sounder from the wheel-house roof above the helm position, but still within reach of the helmsman.

The power supply for the majority of echo sounders will be 12 volt DC and some will also work from a 24 volt DC supply. Several models have a dual 12 or 24 volt capability as well as 9 volts taken from an internal dry battery. These are especially useful for people with inflatables or very small boats that do not regularly have a generated source of power aboard. As with all sensitive electronic equipment a smooth, regulated source of supply is essential for trouble-free operation. If possible, try to ensure the echo sounder receives a single source of power, fused at the boat's main fuse panel.

SUMMARY

- When choosing an echo sounder take into consideration the beam width, pulse rate and frequency of operation, all of which will have a bearing on the performance of the unit.

- The sounder should be fitted within sight of the helmsman which will enable him to use it as an aid to navigation when entering unfamiliar harbours, preparing to anchor or cruising at night.

- Never install the transducer too tightly into its fairing block as this will impair its ability to oscillate and prevent it from giving accurate readings.

11

SPEED AND
DISTANCE LOGS

A speed and distance recorder is an essential part of your navigation equipment. It gives details of how far you have travelled and gives an indication of the boat's speed at any given time. This information is required when calculating a dead reckoning position or a fix. There are several types of log; those which are trailed over the stern of the boat and give a mechanical read-out on a gauge held in the cockpit, those that give an analogue read-out derived from signals sent from a transducer fitted into the hull and the latest digital displays, also worked using pulses sent to the computer from a paddle wheel transducer mounted on the hull. For some smaller craft, a trailing log might be the answer as there is no fitting

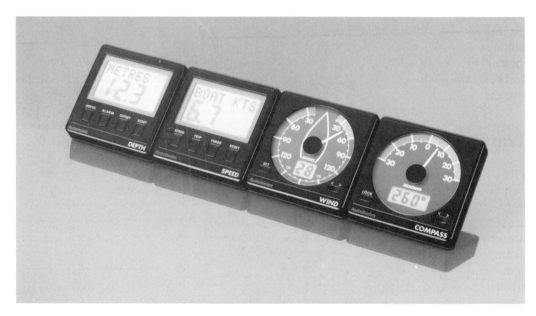

A complete set of instruments for yacht or power boat. From left to right: echo sounder, speed and distance log, wind speed indicator (sailing yachts only) and electronic fluxgate compass. Micro-electronics has made for a small size with many functions and facilities.

procedure. The disadvantage of this type of log is that it has to be unrolled every time you need to know the boat's speed or distance unless you leave it trailing. They also tend to get tangled up on floating or sub-surface debris. For the serious boat owner, who uses his craft on a regular basis, a permanently fitted log is the answer.

Most small pleasure craft usually fit one of the popular through-the-hull-type logs which incorporate an impeller or paddle wheel. The wheel is made up of small blades with a tiny magnet attached. As the wheel spins in the water the magnet actuates a switch which in turn sends electrical pulses up to the microprocessor circuitry in the main display box. These pulses are converted electronically to show speed and distance run. When choosing a log, ensure that you select one with as robust a transducer as possible, and preferably one that can be withdrawn for inspection from the inside of the hull. Bits of seaweed and debris do tend to get themselves wrapped around the wheel and even a small piece of weed can jam the system.

Transducer Siting

The siting of the transducer on the hull is very important. It should not be positioned in an area of turbulence such as that caused by the propeller or other projections on the hull – skegs, P-brackets, skin fittings, A-brackets, the engine, echo sounder transducer and fairing blocks are all items that could cause air bubbles and obscure flow. Small inaccuracies in the hull surface can also cause turbulence and therefore false readings on your log while the friction caused by the water rushing

This simple analogue 'clock face' log is easy to read.

across the hull (skin effect) can also give inaccurate measurements of speed. Fit the transducer in an area which has a smooth flow of water. Unless you know the aquadynamic shape of your particular boat, you may have difficulty in finding such a place, but the boat's manufacturer will have a good idea, so consult him before fitting.

In fast power boats a considerable amount of turbulence can be generated giving rise to all sorts of problems for both log and sounder. A sports boat at speed rarely has the central portion of its hull in contact with the water so siting a transducer here would be folly. The amount of cavitation causes the water in front of the outboard or outdrive leg to be extremely aerated making paddle wheel or pulse

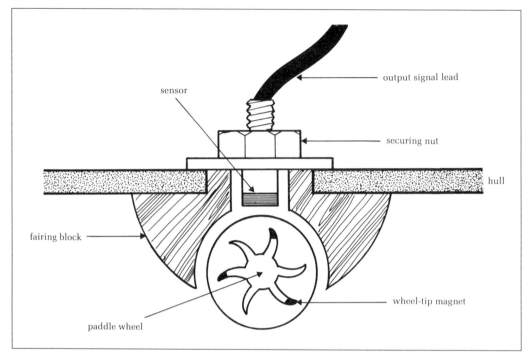

A paddle wheel transducer used on most speed logs. As the wheel turns, small magnets in the blade tips induce a pulse that is picked up by the sensor and converted electronically to give the boat's speed.

generated transducers almost useless at high speed. On a high speed boat it is best to try and site the transducer as low as possible, just aft of the middle third of the hull where turbulence is least vicious.

Fitting the Transducer

The fitting of the transducer itself is not difficult once a suitable site has been selected. With the boat slipped a hole is cut in the hull and the transducer fitted and tightened up onto its rubber sealing gasket using the single or double lock nuts inside. If you fit a retractable type, remember to keep the blanking nut in a

safe place, preferably tied to the transducer casing by a piece of cord. If you have to remove the transducer for cleaning or servicing with the boat afloat, the nut will be required instantly to stop the inflow of water. Never attempt to remove the transducer on your own. If you drop the blanking nut into the bilge out of reach, the consequences could be very serious indeed!

As with the echo sounder, the cable from the transducer to the log display should never be cut short and should be passed through the boat via the shortest route possible using glands and grommets where necessary on bulkheads and partitions. The cable should be kept clear of

The impeller for a speed and distance log mounted in position on the hull of a pleasure yacht. Note the streamlined skeg in front of the impeller which helps to reduce turbulence and ensure accurate readings.

other wiring if possible and should be cleated in the bilge space to a wooden batten and not to the floorboards. Access should be free to all parts of the cable for easy servicing.

Display Unit

Site the display unit in a position that can be clearly seen from the helm. Some navigators like to fit a second log display 'repeater' at the chart table where instant calculations can be done. The supply for the device should, as usual be provided from a separate fused spur from the master fuse panel.

SUMMARY

- A speed log is an essential instrument which is used when calculating your position as well as giving an indication of how far you have travelled.

- Care should be taken to site the paddle wheel transducer away from areas of high turbulence such as near the outboard engine or behind the stern gear support brackets. Choose an area that has a smooth flow of water.

- When removing a transducer for inspection have the blanking nut to hand and do not attempt the operation alone. If you drop the nut out of reach the water flooding in through the hole could soon overwhelm a small boat!

- When fitting a speed and distance log, never cut short the cable which takes the information from the transducer to the electronics at the display. These cables are individually 'tuned' and if cut, will alter the performance and calibration of your set.

- On a high speed boat, such as a small sports boat, it will be best to site the transducer as low as possible at a position just behind the middle third of the hull where the turbulence caused by water flow is at its least vicious.

- Many manufacturers of logs will be able to supply a small repeater unit which comprises a smaller version of the main log display. This is ideal for mounting in secondary positions such as the navigator's chart table or on the flying bridge of a larger cruiser. The units are linked electronically and utilise the same transducer.

12
RADAR SYSTEMS

There are now many types of small boat radar on the market and with prices coming down all the time, more and more yachtsmen are fitting them to their craft. The dangers of being at sea in a thick fog or mist, even if you know where you are on your chart when the fog comes down, is a frightening and very sobering experience which is not to be recommended. The potential of collision with other vessels and unseen rocks, or of going aground in poor visibility is increased and a minute error of calculation can land you, your crew and the boat in a life-and-death situation. This, when considered alongside the vast amounts of sea-going traffic in certain areas, such as the English Channel, makes the proposition of an all-seeing radar set very attractive.

The radar will paint a picture on its screen of everything taking place around the boat in 'real time' and no matter what the prevailing conditions, you will be able to see the position of your boat in relation to other ships and landfalls. The radar is also a very useful navigation tool enabling the user to identify coastlines, harbour entrances, navigation buoys, moored boats and other obstructions. Radar can be particularly valuable when approaching a strange harbour, especially at night, as it allows you to identify piers, jetties and harbour walls as well as the buoyage marking the approach channel. It is also useful when coming in to an anchorage, because by measuring off the distance from your position to two known points on shore, you are able to steer the boat following the range markers on the radar screen until they correspond with the ones plotted, thus ensuring an accurate approach.

Fully Integrated Systems

Electronics has advanced to such a state that the radar can now be integrated with other navigation and position finding equipment like Decca, GPS and Loran. The latitude and longitude is displayed on the radar screen along with the visual picture from the scanner building up a clear and comprehensive navigational picture. This fills the navigator with confidence and, with time, will give him the ability to put more faith into the safe navigation of his boat.

The radar works on a line of sight principle and is made up of four separate sections; transmitter, antenna, receiver and visual display. The antenna and receiver are combined into one unit called the radome which most people think of as the swivelling arm on the outside of the boat. The radome is enclosed in a plastic casing on the smaller pleasure craft radar units and no movement can be seen from the outside. The combination makes installation much simpler. The trans-

The radar comes in a kit comprising scanner unit, mounting bracket, multi-core cable and display/control box.

mitter sends out very short pulses of radio energy at a high frequency and as it rotates it directs these pulses around a 360 degree circle where they are reflected back from any solid object that they strike. The return signals are then electronically processed before being displayed on the screen. Although the pulse is transmitted horizontally, the information when displayed, shows a bird's-eye view of the situation outside the boat. The central point on the display screen is the position of the boat and all calculations are made from this point.

The transmitted pulses are very short in duration because the receiver cannot start to receive them until the transmitter has stopped sending them out. The actual transmit time is measured in microseconds and a radar pulse travels at the rate of one mile every 6.2 microseconds. Typically, the pulse will be $0.1\mu s$ on shorter distance ranges during which time it will have travelled 30 metres. On the longer ranges more power is required to push the pulses out over greater distances.

Compact Units

In times gone by, before the advent of micro-electronics, the radar set was a

Scanners can be mounted in a variety of positions. This one is fitted to the coach roof of a power boat on a sturdy bracket. Note the signal cable is securely clipped down.

bulky and cumbersome device, unlikely to be fitted to any but the largest of boats. These days a small radar unit with a range of about 15 miles can be bought for boats up to about 30 feet in length. The power requirements for this size of set are in the region of 40 watts and will operate from either a 12 or 24 volt DC supply.

Retail prices start at around the £1,000 mark and are very good value indeed. Remember to choose a reputable manufacturer and be prepared to shop around to get the best possible price. Even more money can be saved if the owner fits the radar himself – a task that, after consultation with the manufacturer and supplier,

should be quite within his capabilities.

The radar set consists of a radome or scanner, either mounted in its own GRP encasement inside which it rotates, or free-mounted in the form of a swinging arm; the display screen and controls. The most common type is the enclosed radome which is usually mounted as high up as possible on the boat to maximize performance. This performance can be attenuated considerably by the curvature of the earth. The range of the radar depends upon the correct siting of this radome and if it is mounted too low down on the boat, just because it looks good in that position, serious deficiencies in the range might result.

On many sailing yachts the radar scanner is mounted on a special bracket up the mast. The signal cable is either fed through the mast, clipped down its outside or fed through plastic conduit to prevent tangling in rigging or sails.

Mounting the Scanner

There are many brackets, rigs and short mast systems available to enable you to mount your scanner in the appropriate location on your particular boat and the manufacturer or chandler should be able to discuss with you the best bracket for your needs. Another reason for mounting the radome as high as possible is for safety purposes. Radars scan using microwaves similar to those generated by a domestic microwave oven. It would be foolish to mount your scanner at a height where you or a member of your crew could look directly at it when it is switched on. As radar works roughly on a line of sight basis, generally the rule is the higher the scanner, the greater the range, depending upon the output power of the set.

The radome should also have a clear 360 degree view and should not be mounted in a position where part of the boat will obscure that view. The last thing to ensure is that the scanner itself is mounted in a fore-and-aft line with the boat which means that the ships heading marker on the radar screen will be true with the scanner.

The radome scanner is coupled to the

The radar multi-core cable should be fitted with a waterproof deck plug and socket at the point at which it passes through the deck.

main radar by a thickish multi-core cable. Usually a coil is provided with the unit which is of sufficient length to allow mounting of the scanner and radar set in a wide variety of boats. Routing this main cable is just as important as the siting of the scanner and some thought should be given to this before starting work.

One of the first things to do is to connect the one end of the cable into the radome following the manufacturer's instructions, securing it in the special clamp and running it through the waterproof grommet as it passes through the GRP casing. This is much easier to do in the comfort of the cabin than half-way up the mast or balanced on the cabin roof. The cable should pass through from the outside to the inside of the boat via a proper waterproof deck gland or bulkhead coupling. A hole bunged up with mastic is just not good enough. If the cable is up a

stay sail mast such as those found on some fishing cruisers or on a motor sailer, it should be either threaded through the hollow mast section or led down the outside and protected by a piece of electrical conduit.

Never run the cable in such a way that it will get snagged by boots or deck gear or across areas which could cause chafing or cuts. Also, do not allow the cable to be kinked sharply which might damage some of the internal small gauge wires.

The terminal end of the cable should be supplied with a special plug which fits into the rear of the radar unit. Remember to route this well away from sources of possible interference such as compass, echo sounder and ignition circuits. It should also be kept clear of any areas of heat. Always remember that taking a little time and trouble at this stage will prevent possible equipment failure due to corrosion and water ingress in the future.

Mounting the Rig

The radar models available are all basically the same shape and physical size and can usually be mounted in a number of different ways, from slinging them on overhead brackets, to mounting them in dash consoles or by simply siting them on a flat table surface at the helm or near the chart table and navigator's corner.

Obviously, the screen and controls need to be visible from the helm and should be shaded if possible to prevent direct sunlight from blanking out the picture. Special shades are available with some models or you could try making your own from sheets of black flexible plastic.

Apart from mounting the bracket, bolting the display down to it and connecting up the scanner cable as previously described, the only remaining connection is the power supply. This should be wired in using a suitable gauge of cable (once again see manufacturer's instructions) and should be run in as short a route as possible from a fuse or circuit breaker at the boat's main distribution board. The handbook supplied will give the fuse rating and you will probably find that a roll of wire and an in-line fuse and holder are also supplied with the radar.

Some of the radars available today will require a degree of fine tuning in order to match them to their installed conditions and to ensure maximum working efficiency. Each one is individual in this respect, but usually the details supplied on tuning are sufficiently simple enough

for the competent DIY owner to understand them. It should be remembered that before starting a DIY installation of a radar, check with your supplier about any invalidity of the guarantee if you fit it yourself. Some manufacturers might insist on an independent check of the installation before commissioning to ensure that it has been carried out correctly.

The possibility of self-installation might well save you money and it can give some boat owners much satisfaction from seeing a good job well done. It really is up to the individual and his feelings as to whether he possesses the necessary skills to carry out the task. Remember that the operating efficiency of a radar set will only be as good as its installation on the boat.

SUMMARY

• Mount the radar scanner in a position as high as possible on the boat to get the maximum range performance. Never mount it in a position where it can be directly looked at as the high frequency microwaves could be dangerous.

• When installing the cable from the scanner down a tall mast, always remember to clip it up at regular intervals or better still run it in conduit. This will prevent it snagging on rigging and sails.

• When siting the radar display unit, position it well away from the ship's compass. If it is too close, it may upset the compass calibration, throwing out your navigation calculations.

• The advances in electronics now means that the radar system can be fully integrated with other navigation electronics such as the echo sounder, log, Decca navigator, GPS satellite positioning systems and Loran.

• Never run the multicore cable from the scanner to the screen display in such a way as to allow it to become snagged on deck by rigging, boots or deck gear. It should be taken through the deck to the inside of the cabin as soon as is practically possible, using a waterproof gland – preferably a plug and socket type.

13
WEATHER
INFORMATION

The Navtex weather and navigation information system has been operating in Northern Europe for several years now and is a free service which will ultimately provide worldwide coverage giving navigational, meteorological and distress messages. The messages are in English, although some countries send their information in a combination of languages, and is paid for by the particular government of the country and in the UK is provided by BT.

The primary reason for carrying a Navtex is for the weather. A full weather report is given at least twice daily and gale warnings are displayed as soon as they are issued by the meteorological office. For the cruising yachtsman the navigational warnings are also of great benefit. These will report if a buoy or lighthouse is not flashing, where cable laying is going on, if an oil rig is being towed slowly along and other hazards. There are also ice reports, search and rescue information, pilot messages and information of Loran, Omega, Decca and Satnav services.

Video Display

The early units incorporated a printer which printed onto cash roll-type paper. This made them very expensive and cumbersome, as both receiver and printer were required. The second generation of Navtex receivers utilizes a digital or video display which does away with the need for awkward paper print-outs. Several incorporate a powerful memory bank which can store several 'pages' of information and which can be scrolled through and easily read. Not having a printer attached, these units are much cheaper than the older types.

It is not unusual to receive stations over 600 nautical miles away with just a deck-mounted antenna, in good conditions. The system operates on 518kHz and units are supplied with an active antenna. Due to the low frequency nature of the system only deck-level antenna mounting is required. The memory can store about 3 days' information from any one station, depending on the volume of messages. In fact, one problem with the system is the quantity of messages transmitted! To accommodate this the receivers are computer controlled and can be programmed to receive only the stations and message types required.

The past few years have seen a considerable growth in the Navtex service with many more stations becoming operational. Of particular interest, the American East

An example of a weather facsimile receiver which produces weather and navigational reports on a roll of paper. Other types have the information displayed on a cathode ray screen.

Coast is now covered, as is the Western European Seaboard (including the Azores) and the Mediterranean. With regard to the Mediterranean service, The Western Mediterranean is well covered apart from a small 'blind' spot around Italy, and there are stations now fully operational in Spain. Continual development will mean more stations and better coverage in the future.

Fitting a Unit to the Boat

The fitting of a Navtex unit is really quite straightforward. The unit comprises an aerial with a set length of cable and a receiving unit. The normal power requirements are a supply up to 14.5 volts DC and the unit should be fed by a separately fused circuit. The aerial cable will be long

111

enough to allow the aerial to be sited on deck and should not be cut as it will have been 'tuned' to the set in a similar way that a VHF radio-telephone aerial cable is. Site the aerial on deck, perhaps on the wheel- house roof or at the top of your goalpost mast or flag mast. Once all connections have been made, all that remains is for the manufacturer's instructions to be con- sulted before switching on.

SUMMARY

- A weather fax machine or Navtex can be useful not only for gale forecasts but also for hazard warnings such as buoys off station, inoperative lights, drifting vessels and towed oil rigs.

- Many of the Navtex units now available have a memory that will store several 'pages' of information for recall at a later time.

- Mount the receiving aerial as high as possible.

This earlier type of Navtex unit from Lo-Kata uses a paper roll which can be handy if you want to file your weather reports and changes to navigational marks etc, prior to updating your charts.

14

IN-BOAT
ENTERTAINMENT

Boating today is a far cry from boating of old where basic facilities and a distinct lack of comfort were the order of the day. Now, many people want a standard of living afloat similar to that which they enjoy at home and this, unfortunately perhaps, includes a full range of entertainments from a simple radio/cassette to a sophisticated, remote-controlled colour television and video recorder.

The choice of systems to fit are as vast as your imagination and with prices to suit most pockets. It really depends on what home comforts you wish to have duplicated on board your boat. If you own only a small craft you might just opt for a portable radio/cassette or something really simple like a personal stereo. If you own a larger boat with a cabin attached you may want to fit a cassette player, radio and small television.

Power Supplies

One of the points which will determine the type of system you can fit to the boat will be the power supply. Obviously if you opt for a portable set, the batteries are self-contained. A radio cassette player is more than likely to be the sort that has been designed for the car and will therefore operate from a 12 volt DC supply, which, of course, is ideal for most boat electrical systems.

Obviously for a high current device like a television, you will need a 240 volt supply or an inverter which converts the 12 volts DC to 240 volts AC. These are available in a range of outputs and cost upwards of £170. They are quite good, but do not like being overloaded. The better ones feature a trip mechanism to protect them from overload conditions. The other answer is to buy a fixed or portable generator to provide your mains supply and this is discussed in full in Chapter 6.

Choice of Sets

The choice of radio/cassettes, as we have said, is very wide indeed. Drop into your local high street electrical retailers or car accessory shop to feast your eyes and to make your selection. Prices start as low as £15 and for this you can expect a basic radio, probably LW/MW/FM, with an output of about 8–10 watts. It will be complete with a below-average speaker and coupling cables. This sort of unit would be ideal for a small day boat or where budget is a prime consideration.

Next up in the range would be a radio

The standard component parts of a car-type radio cassette player including electrically operated aerial and switch.

unit with integral cassette player. The cost will start at around the £35 mark and the unit would be adequate but would not break any records when it came to sound reproduction! These are adequate for a fast, open sportsboat or a small week-ender.

For bigger boats, the choice of quality equipment is better, with some companies producing good value sets which incorporate a stereo FM radio and first-rate cassette player. They start at about £50 and come with twin high-quality speakers, power cables and full fitting instructions. These units, when installed in a boat cabin or in the stern cockpit of a smart sports cruiser can really enhance anybody's cruise. Top prices for this sort of set would be in the region of £100–£180.

Television Receivers

The choice of television sets is slightly more restricted, but you will still find a good selection at your local television shop. The ideal size of screen is the 'portable' size of around 30–36cm (12–14in) measured diagonally across the screen face. You can sometimes pick one up second-hand, but beware because, the tube might be worn out and you might end

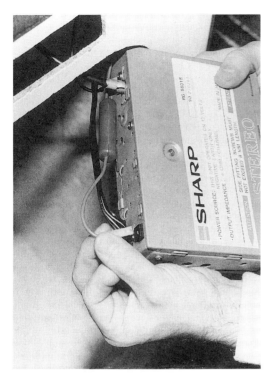

This radio is being fitted into the top of a small cupboard on a boat. Power and aerial connections can be seen.

up paying for something that will only last you three months without having the security of a guarantee.

High street stores will sell you a reasonably good value 36cm (14in) set for around £129 for a non-remote control model, or a 36cm remote for £160. For those who don't want colour there are still a few mono sets available from various outlets with prices starting at about £49.

Video Systems

When it comes to picking a video for the boat your choice depends upon whether you want to just play back pre-recorded tapes or record material from the television and play this back. Whichever system you go for you will need to have either a generator or an inverter because you will need the full 240 volts AC for operation. Video players only are available and are similarly priced to the smaller colour televisions at about £160. Many have all the basic functions of a full-sized model; play, fast forward, rewind and picture search, but the overall size is about half the width of a standard video recorder, making it easier to fit into the confined space of the cabin.

Full-size machines can be bought with all-singing, all-dancing functions right up to several on-off settings that can be programmed in for anything up to a year! Prices vary from £269 up to about £700 for full functions and NICAM digital stereo.

Video players, cassettes and televisions are by their very nature sophisticated pieces of electronic equipment which react very badly to a damp atmosphere. Couple this with the watery environment of a boat and you have a dangerous cocktail which could ruin hundreds of pounds worth of entertainment gear. I once saw a popular make of open sports cruiser with a cassette radio fitted in a position on the dash which, when cornering in a bit of a chop, took the full force of side slip spray. Needless to say, it did not last five minutes.

Siting the Equipment

Protection for cockpit sets is available in the form of plastic covers that bolt into place under the front escutcheon plate of the radio and feature a hinge-down face that can be lowered to put in a cassette or turn on the radio, and closed when under-way. They are available from some chandlers or you could have a go at making your own out of perspex sheet, plastic glue and stainless steel hinges.

Try to site your equipment out of the direct line of spray or water coming aboard. If the boat has a cabin, fit the unit inside and install waterproof marine-type speakers in the cockpit where you want to use them. Special 'marine quality' speakers can be bought which are totally waterproof and ideal for the environment of a boat.

Televisions and video players should be well secured in their mounts to prevent possible shifting in a heavy swell or at speed. The best way is to enclose them in a locker after you have removed the front, or in a cupboard where the door can be used to close the television away when not in use. In any case, bolt them down or stow them safely before starting off.

Aerials

Obviously you will need an aerial for the radio or television set. Several are available from long whips to those small, rubber types that flex if you crash into them. These are a good idea on a small boat. Mount the aerial as high as possible for maximum reception and don't forget that if the cable has to pass inside the boat to use a waterproof deck gland or rubber boot. This also goes for any power leads you might want to route.

There are several types of television aerial available covering a wide range of styles and prices. Omnidirectional ones which cover the entire frequency band-widths are, in my experience very poor compromises. Over the last 20 years of television boating I have come to the conclusion that a simple, cheap set-top aerial costing about £8 does the best job. Buy one that is labelled as multi-band and you will be able to use it in any area. If the local transmitter is vertically polarized you can simply turn the aerial around to compensate.

Boosters are available for areas of weak signal strength and a special 12 volt version for boats can be bought.

SUMMARY

- If you want to install a colour television or video recorder aboard you will need either an inverter that will convert the 12 volts DC to 240 volts AC, or a generator.

- Remember to fit radios well away from areas where they may get wet. This is especially important on small, open sports boats. If possible fit them in the cabin. Waterproof speakers for marine use can be bought for installation in open cockpits.

- Try to buy a radio/cassette that can be removed from the boat when not in use. If you have to leave items such as television and video aboard, lock them up in a secure cupboard.

15
SIMPLE
FAULTFINDING

If you have fitted out your boat from a bare hull, you should have retained your initial plans and wiring diagram and this will be of great assistance in helping you to trace any faults that might occur. Keep your plans safe, preferably by photocopying them and keeping one set at home and another aboard the boat in a waterproof plastic wallet alongside your charts and almanacs.

What do you do if you have no circuit diagram or have just bought the boat? It can be a very frustrating experience trying to trace faults in the fabric of a boat's electrical system. The best method is to apply a modicum of logic to the problem, and success will usually follow. The first step is to try to identify the general area of the fault and only then should an attempt be made to trace it.

Charging Faults

If the fault seems to be in the charging circuit, check your ammeter on the main engine control panel. A constant discharge when the engine is running (and charging should be taking place) indicates that the alternator is not feeding the battery and you will need to look at the circuits relating to this. These include the alternator itself, the blocking diodes or electronic charging sensor and the solenoid circuit or associated wiring.

If, for example, the VHF radio was 'dead', the fault could be either the lack of a power source or a fault with the radio itself. The first of these will be fairly simple to detect. Firstly check the fuse at the back of the set by removing it from its screw-in holder, if this is OK, check the fuse feeding the circuit for the radio at the main distribution board. If this too is intact, and the radio still has no power, the fault obviously lies in the wiring between fuse board and radio.

Test Equipment

After you have tried the simple things you may need to use some test equipment to help you further. There are two types of test gear worth carrying on board to check out electrical circuits and equipment. The first is a simple test lamp comprising a bulb holder on a length of solid-cored cable to give it some rigidity in use. A 12 volt bulb is fitted to the holder and the lamp can be used to check for the presence of voltage at various points in the circuit – for example, across the terminals of a light fitting. This method can be used, with

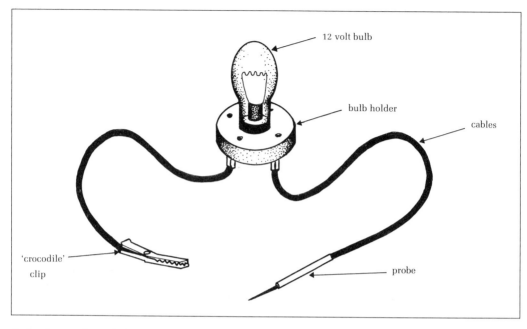

A simple 12 volt test lamp made up from a car headlamp bulb and holder, flexible cables, a
probe and crocodile clip can be very useful in tracing circuit faults.

care, on the low voltage circuits but
should never be used on any high voltage,
mains or shore power circuits. It should
also be remembered that the action of
putting the lamp across the terminals may
cause sparks to be generated.

The other item is a small multi-meter
which should always be kept with the
boat's toolkit (see Chapter 16) and is a
most useful tool for determining the
presence of voltage and for ascertaining
continuity of cables. They are reasonably
priced and one which will read AC voltage
and current up to about 500 volts, DC volts
up to 50 and which also has a resistance or
ohms (Ω) range for continuity measurements should cost you no more than about
£20 from shops such as Tandy or Laskys.

Use the meter on the DC current setting
to measure your power. The meter will
have two probes attached to two cables,
one red and one black. The black one
should be attached to the negative power
lead and the red to the positive. A voltage
reading on the meter face should be
between 12 and 14 volts DC. If no voltage
appears at this point, work back down the
line until you locate it. The problem could
be a faulty fuse holder or broken cable.
Switch the meter to ohms to measure
across a single cable (after the power has
been isolated) which will determine a
break in the cable's continuity. A break
could also be found by testing with the
power switched on, but you would not
know if the supply or return wire was at
fault. Your meter can of course be used to
check fuses, bulbs or fittings on the
resistance range, a high resistance indicating that a replacement is needed.

With a simple case such as a light which does not work, the first area to look at is the bulb itself by removing it and replacing with a new one from your box of spares. If this still doesn't work, you have narrowed the problem down to the circuit for that particular light or the fitting. Do any of the other bulbs on the same circuit light up, or is this light the only one affected? Once again, check the fuse or circuit breaker for that particular run. This sort of logical approach helps you to narrow down the area at fault, but may not actually locate it unless the bulb or fuse has blown. This is where you can employ your test equipment.

New Cables

By logical use of the test lamp or meter, wiring faults or even faults in fittings should be easy to locate. Replacement or repair of the fitting is usually relatively simple and a small selection of lamp holders, bulbs, fuses, cable and tape should be carried in the electrical tool-box. Running new cabling through the boat can be a bit more difficult and it is always a wise precaution to run your cables through conduits wherever possible. These wires will then be instantly accessible as and when you need to check for faults or to replace when worn. A judicious use of colour coding will also pay dividends when it comes to cable tracing, especially if the various cable colours are noted on your master wiring plan.

Faults in your engine starter circuit can be more difficult to locate. The solenoid circuits should be checked in the same way as other normal circuitry. The heavy-duty cables used in the starter circuit may show good continuity when tested with

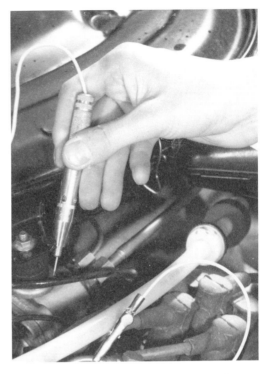

Using a low voltage circuit tester to trace faults on the ignition system of a petrol engine.

the lamp or meter, yet still not be capable of passing the heavy currents required for the starter motor without a large drop in voltage. You will know if the solenoid is working as it will make a distinct clicking noise when the key is turned in the starter switch. If all checks here are okay, carefully examine all the connections on the starter motor and associated cables, making sure that good, tight, dirt-free connections are being made at these points.

Corrosion on the main battery terminals is one common fault here, which can be a prime cause of your voltage drop in the circuit. Make sure that they are cleaned at regular intervals, tightened and coated

Two examples of terminal and wiring kits. On the left, a special crimping tool and various cable terminations, to the right, terminations, wire cutters and electrical screwdrivers.

with a marine electrical waterproof grease. If everything else has been checked and the engine still refuses to turn over, or does so very slowly, suspect the starter motor itself. The starter motor can usually be taken off the engine without too much effort and is serviced in a similar way to that of a car type.

These motors are usually so well built and of such a sturdy construction with great inherent reliability, that it is very unlikely that a unit will need dismantling until totally worn out and in need of replacement as a whole. However, stuck shafts, broken solenoids or worn armature brushes can usually be repaired by the amateur. If in doubt, check with the engine manufacturer or employ a qualified marine electrician to do the work for you. Faults in the starter circuit can be some of the worst to find and an obvious first step would be to check the main drive belt from engine to alternator to ensure that power is actually being generated. Next, check out the wiring as far as you can for any obvious breaks, poor connections or corrosion. The charging light on the engine warning panel is a sign that things may not be all they should be. If you have a voltage regulator or alternator problem it is usually best to call in an expert.

A simple electrical toolkit comprising

your test lamp, multi-meter, spare fuses, bulbs, connectors, wire as well as tools such as wire strippers/cutters, small screwdrivers, long nose pliers and insulating tape, should form part of the extensive toolkit that should be kept aboard every sea-going boat. We tend to rely heavily on our electrical systems, and a failure at sea could have serious consequences. Having the tools to help locate faults and effect a repair, however temporary, will give peace of mind and reduce the anxiety of a failure.

SUMMARY

- If the fault seems to be in the charging circuit a good indication that something is amiss can be seen by studying the ammeter. A constant discharge when the engine is running indicates a possible fault with the alternator.

- One of the most useful items of test equipment is the multi-meter. This can be used to check for voltage, current and continuity of cables.

- One of the main causes of electrical failure is badly corroded or loose battery leads. These should be regularly checked, tightened and greased.

Most electrical faults on open sports boats are caused by spray and salt water getting into the instruments. A possible answer to this problem is provided here by encapsulating the boat's Decca navigator unit in a tough, waterproof box.

16

THE TOOLKIT

As Chapter 15 on faultfinding explained, there will always come a time when you will need to make repairs to the boat's systems and services while at sea. This might include repairs to the electrical system, but could well involve repairs to a faulty toilet, engine or part of the boat's superstructure or hull. In any of these circumstances it is quite likely that some tools will be needed to help effect a decent repair. Many boat owners prefer to carry a full kit of tools with them at all times, not only for the routine painting and repair jobs mentioned above, but also as a vital means of averting danger if the engine breaks down or an important item such as a steering cable breaks when in tidal waters.

Having the right tool to hand is, or should be, part and parcel of good ship husbandry and in a crisis could make a crucial difference in a life-and-death situation. This book concerns itself mainly with electronics and electrical apparatus but in the interests of completeness, I shall describe all the tools necessary to comprise a full and functional boat toolkit. For repairs and maintenance, the tools required depend largely on the type of work to be carried out, your own skills, the complexity of any work you are likely to attempt and the capacity of the boat's lockers to store them. When choosing a set of tools, buy the very best that you can afford, as it will be well worth it for reliability and a successful repair.

Looking After Your Tools

Tools must be looked after properly – brushes should be cleaned out after use, sockets returned to their set and glass fibre resins cleaned from palette knives and mixing rods. Care of your tools is an on-going process, with a major check and replacement of worn or broken implements during the spring refit when many will be required for preparing the boat for cruising.

The tools should be stored in their own watertight boxes to protect them from the inevitable dampness of their surroundings and to help minimize rusting. There can be nothing more frustrating than a pair of pliers jammed shut with rust! The typical metal tool-boxes used by mechanics are hardly suitable for life afloat as they hold water and will actually increase the chances of the tools inside rusting.

Ex-Army ammunition boxes are more suitable as they incorporate a rubber gasket seal in the lid. They are strong and can be obtained reasonably cheaply from government surplus stores and will stack for ease of stowage. Other suitable containers are large plastic lunch-type boxes for items such as glasspaper and paintbrushes, electrical cable and fuses. A sackcloth plumber's tool bag which is strong and does not rot is ideal for oily spanners, filters and bits of engine.

There are five different categories of

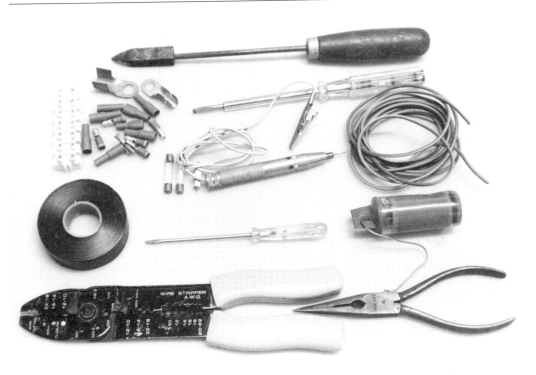

The basic electrical toolkit comprising spare wire, circuit tester, pointed nose pliers, crimping tool, bullet and spade terminals, fuses, PVC tape, barrier strip, solder and a soldering iron that can be heated up on the gas ring of the boat's cooker.

tools for repair and maintenance: electrics – wiring, batteries, fuses, switches and navigation and cabin lighting; glass fibre maintenance and repair; painting; woodwork and mechanics – engine servicing and repair, water and heating systems, plumbing, services, pipework are running and standing rigging. Many of the tools will be common to two or more groups, but it will be best to sit down with a pen and pad when compiling your kit and list all the tools you think you may need in each category, then hone them down to the minimum required to adequately cover each job. This prevents doubling up and will not only reduce the amount of storage space needed but will also of course, save money.

Tools for the Electrics

To assist with the care and repair of electrics and electronics afloat, a small set of terminal screwdrivers (both cross-head and ordinary blade) will be required. A pair of snips for cutting and trimming electrical wiring are also essential. For those wishing to make joints in their cables or to produce good, trouble-free terminations, a special crimping toolkit with various bullet or clamp terminations would be a boon. A low wattage 12 volt soldering iron could also come in handy when it comes to sweating a joint or making an emergency repair to a damaged printed circuit board, but for safety's sake a soldering iron that can be heated up on the cooker hob should be included in case the boat's main supply is inoperative.

Other items should include various rolls of tape; ordinary PVC electrician's tape as well as the excellent self-amalgamating tape which is extremely waterproof and can be used for everything from emergency cable jointing to sealing junction boxes and lids. A tube of silicone rubber sealant of a type that can be applied from a 'gun' is useful for sealing holes where cables have been passed through decks or cabin sides, and a small tube of silicone grease and a can of WD40 will keep battery terminals clean and switch contacts free from corrosion.

A waterproof box containing spare fuses for the boat's various systems and equipment should also be part of the electrical tool-box. Fuses should be a mixture of straight and quick blow and at least three of each type should be carried. A couple of spare circuit breakers, although expensive, will give peace of mind if you have this sort of fusing system fitted.

Tools for Glass Fibre Repair

The care and repair of glass fibre requires no special tools and the kit can be made up of the following items: resin, matting in the form of short strips, glass-fibre tape and catalyst. For minor damage, scratches and blemishes, a quantity of gel coat touch-up paste in the correct colour should be included along with several grades of wet and dry papers. For more serious damage a filler kit, such as those that can be purchased from most motoring accessory shops (they will be cheaper than the same thing bought in a chandlers), can be used in an emergency. Other items might include polythene sheeting, to assist curing and give a smooth finish, mixing spills and paper cups for the resin.

A palette knife and a packet of cheap plastic gloves will also be useful to protect the hands when handling resin. These materials, like the turpentine, should be carefully stored as the resins and hardeners are flammable. Mark on the packs the date of purchase, as items like resin have a limited shelf life.

Tools for Paint Jobs

One of the most important aspects of painting is achieving a good finish, so tools for preparation, such as a variety of scrapers, glasspaper and a can of paint remover, should be included along with a good wire brush to help remove flaking or chipped paint.

The number of brushes you require will depend on the type and area of the work you intend to carry out. As a minimum I would say one 6in, 3in, 2in and 1in brush together with a couple of artist's or signwriter's brushes for those awkward areas.

Some turpentine is also useful for cleaning (watch where you store it) and some mild detergent to finish off the process. Keep an old rag with the kit and a pair of rubber gloves and an overall which can be used for all the DIY jobs.

Probably the best method of storing painting items is in a large plastic bucket which will keep them together and also allow air to circulate.

Tools for Carpentry Work

Many boat owners will have to do some carpentry from time to time as a good deal of the boat's interior will have been made from timber. Basic woodworking tools for use afloat include screwdrivers of both types, a small padsaw, tenon saw, hammer, pliers, files (round and flat section), a sharp craft-type knife, glass-paper, nails, screws and a small tube of ready-mixed glue.

For outside varnishing and repair the kit might also contain one or two small tubes of wood stopper and a clear exterior-grade varnish.

A bradawl or auger would be useful for starting screws and a small hand drill and selection of bits would complete a very comprehensive set. Don't forget such things as marking pencils and a steel roll-up tape measure.

Tools for Mechanical Jobs

Next to the electrical set, the most important set of tools on the boat will be those used to repair or service the main engine. The range required will depend largely on the engine, its type and installation, but with an outboard motor a plug spanner, emery paper and flywheel pulley should be carried, along with spare points, a feeler gauge to set the points and the plug, and various screwdrivers to make any adjustments.

For inboard engines or outdrive legs, you need a small socket set plus open-ended and ring spanners in the same sizes or larger. A wooden mallet, cold chisels and tin snips which also double as wire cutters would be useful for the propeller in the event of a serious tangle, while a pair of clamp grips or an adjustable spanner are useful for removing stubborn nuts or acting as a third hand.

Finally, it is worth noting that some items, although they may only be required rarely, are useful to have aboard. Whipping twine, self-amalgamating tape for sealing rope ends and a pointed, hollow metal spike (Swedish fid) for splices will help you to keep your ropes and lines in trim. If you have a wooden hull or planked deck, caulking irons can be made from old cold chisels, reshaped and ground down, and how about a powerful recovery magnet to look for those lost boat keys or other tools dropped overboard?

SUMMARY

- A range of tools are a must aboard any boat. Before you buy, take time to list those tools that will be of most use to your type of craft and the areas you will be working in.

- Metal toolboxes are a bad idea in the salt-laden environment of a boat. Much better are ex-Army ammo boxes which have a rubber gasket sealing the lid. Alternatives are plastic lunch boxes or folding 'tiered' tool-boxes.

- Remember to keep a box of spare fuses, cables, tape, light bulbs for cabin and navigation lights and various cable connectors alongside your regular tools.

GLOSSARY

Alternator A device which generates current and voltage for battery-charging purposes.

Batteries The storage battery is an electro-mechanical device which converts chemical energy into electrical energy and stores it ready for use. Batteries need regular recharging after a period of use. This is normally done by connecting an external DC current across the terminals, usually supplied from an engine-driven alternator.

Batteries – maintenance-free Special marine-type batteries that are sealed and have integral carrying handles fitted. They do not require attention other than to maintain the state of charge and are safe to use in the environment of a boat.

Batteries – twin On most marine electrical installations, it is common practice to fit two batteries, one for engine starting, the other to supply current for the boat's domestic facilities (the lights, the radio and so on). The batteries are charged from the same alternator with the charge current split between each battery by using a blocking diode.

Double wiring system All circuits should be fitted with a double wiring system comprising separate cables for both outward and return legs. This helps to prevent corrosion and current leakage.

Echo sounder A useful device using pulses of energy, sent out from a transducer below the boat, which bounce back and are converted electronically into a readable form to indicate the depth of the water beneath the hull.

Fuses Fuses come in all shapes and sizes, from the familiar cartridge type to more sophisticated trip devices. It is essential that all circuits in the boat's system are fitted with a correctly rated fuse.

Galvanic corrosion This is caused by small electrical currents flowing between dissimilar metal parts of the boat that are immersed in sea water, for example, a mild-steel rudder plate fitted to the boat with bronze pintles and brass screws.

Generators – fixed These are installed permanently inside the boat and can be chosen to provide a 240-volt AC 'mains' output which will run colour televisions, hairdriers, air-conditioning plants and tools. They are quite large, necessitating a spacious boat.

Generators – portable A portable generator, like its fixed big brother, is capable of supplying 240 volts AC in a range from about 350 watts to 1,500 watts. They are much easier to handle, can be used outside the boat and are cheaper than fixed models.

Isolating switches These are fitted into the positive line from the battery to the distribution board or fuse panel. They are an essential safety item as they can be used to isolate the electrical supply from the circuits in the boat in the event of a fire or other major fault occuring.

Log The speed and distance log is a recorder which shows how far the boat has travelled and how fast it is moving through the water. It is a very useful item for helping with navigation calculations.

Navigation lights All craft that cruise at sea after dark should be fitted with navigation lights. The sizes and fitting criteria are laid out in the *International Regulations for Preventing Collision at Sea* (I.M.O.).

Navtex Navtex is a weather and navigational information gatherer that receives information from various transmitters around the coast and converts it into readable pages on a video screen. The service is regularly updated and pages can be stored in the device's memory for recall at a later time.

Night vision At night, it is important for the helmsman to keep his 'night vision', which can be ruined if beams from the navigation lights shine in his eyes. It is therefore preferable to fit shades to those portions of the lights that shine directly on to the helm position, remembering that no portion of the shade must obscure the particular lights' designated angle of view.

Radar Small boat radar sets are now freely available for the average boat owner. A useful tool, they give the helmsman a visual picture, on a screen, of objects (for example, other boats, the coastline and so on) that surround his craft. Series of high-energy pulses are sent out from a special scanner mounted as high as possible on the boat. These pulses are reflected back from distant objects where they are converted electronically into a visual interpretation of the surroundings.

Resistance A resistance in a circuit can be likened to a sort of brake which slows down the stream of electrons passing through the circuit. It is measured in Ohms and the formula for calculating it is I (current) equals V (voltage) divided by R (resistance).

RFI (Radio Frequency Interference) This is caused by the boat's electronics being acted upon by windscreen wiper motors, petrol ignition systems and other similar electronic activity. A crackling, fizzing noise can be heard over the VHF or other equipment when unsuppressed circuits are feeding this form of interference. RFI can be eradicated by the judicious use of suppression capacitors which are fitted to motors, ignition, circuits and so on. Screening and bonding of cables can also help.

Sacrificial anodes These are used to overcome the problems of galvanic corrosion. Usually made from zinc, they are fitted to those exposed metal parts of the boat (hull, rudder, stern gear) that are susceptible to corrosion. Being much lower down the galvanic scale, they corrode harmlessly away instead of the metal parts of the boat.

VHF radio This should be standard equipment on any boat that goes to sea as it is an excellent means of communicating with other vessels or shore-based establishments such as the coastguard or the marina. It is also the very best way of calling for help in an emergency. A wide range is available to suit most boats.

INDEX